Renegotiating Authority in EU Energy and Climate Policy

T0260355

In the context of multiple crises, EU Energy and Climate policy is often identified as one of the few areas still exhibiting strong integration dynamics. However, this domain is not exempt from contestation and re-nationalization pressures. This collection seeks to understand those contradictory integration and disintegration tendencies by problematizing the notion of authority: When, why, and by whom is EU authority in Energy and Climate policy conferred and contested? What strategies are used to manage authority conflicts and to what effect? These questions are examined in some of the knottiest aspects of EU energy and climate policy, for example, the adoption of the landmark Governance of the Energy Union Regulation, the long-drawn-out attempts to complete the EU's internal energy market, the struggle to achieve ambitious EU targets in renewable energy and energy efficiency beyond 2020, the blurring of economic and security instruments in external energy policy, or the heated discussions over the Nord Stream 2 gas pipeline.

The chapters in this book were originally published as a special issue of the *Journal of European Integration*.

Anna Herranz-Surrallés is Associate Professor of International Relations at the Faculty of Arts and Social Sciences, Maastricht University, the Netherlands. She specialises in EU external energy policy and foreign investment governance, with a focus on the nexus between security, sustainability and democracy.

Israel Solorio is Associate Professor at the School of Political and Social Sciences, National Autonomous University of Mexico. His research focuses on the interlinkage between climate and energy policies, having expertise on energy transitions, the promotion of renewable energy and the democratization of energy, policy integration and national climate policies and socio-environmental conflicts around energy projects.

Jenny Fairbrass is Associate Professor of Business and Management at Norwich Business School, University of East Anglia, UK. Jenny's research primarily revolves around EU public-policy and policy-making processes with a focus on the interface between sustainability, environmental, and energy policy as well as Corporate Social Responsibility (CSR) and Business Ethics.

Journal of European Integration Special Issues

Series editors:
Thomas Christiansen, Maastricht University, Netherlands
Simon Duke, Formerly at European Institute of Public Administration, Netherlands

The *Journal of European Integration* book series is designed to make our Special Issues accessible to a wider audience. All of the themes covered by our Special Issues and the series are carefully selected with regard to the topicality of the questions addressed in the individual volumes, as well as to the quality of the contributions. The result is a series of books that are sufficiently short to appeal to the curious reader, but that also offer ample depth of analysis to appeal to the more specialist reader, with contributions from leading academics

Responses to the 'Arabellions'
The EU in Comparative Perspective
Edited by Tanja Börzel, Assem Dandashly and Thomas Risse

Representation and Democracy in the EU
Does one come at the expense of the other?
Edited by Richard Bellamy and Sandra Kröger

Coping with Crisis: Europe's Challenges and Strategies
Edited by Jale Tosun, Anne Wetzel and Galina Zapyanova

Globalization and EU Competition Policy
Edited by Umut Aydin and Kenneth Thomas

Redefining European Economic Governance
Edited by Michele Chang, Georg Menz and Mitchell P. Smith

Understanding Conflicts of Sovereignty in the EU
Edited by Nathalie Brack, Ramona Coman and Amandine Crespy

Economic and Monetary Union at Twenty
A Stocktaking of a Tumultuous Second Decade
Edited by David Howarth and Amy Verdun

Renegotiating Authority in EU Energy and Climate Policy
Edited by Anna Herranz-Surrallés, Israel Solorio and Jenny Fairbrass

For more information about this series, please visit: www.routledge.com/Journal-of-European-Integration-Special-Issues/book-series/EUI

Renegotiating Authority in EU Energy and Climate Policy

Edited by
**Anna Herranz-Surrallés, Israel Solorio
and Jenny Fairbrass**

Routledge
Taylor & Francis Group

LONDON AND NEW YORK

First published 2022
by Routledge
2 Park Square, Milton Park, Abingdon, Oxon, OX14 4RN

and by Routledge
605 Third Avenue, New York, NY 10158

Routledge is an imprint of the Taylor & Francis Group, an informa business

© 2022 Taylor & Francis

British Library Cataloguing-in-Publication Data
A catalogue record for this book is available from the British Library

ISBN13: 978-1-032-00170-8 (hbk)
ISBN13: 978-1-032-00176-0 (pbk)
ISBN13: 978-1-003-17310-6 (ebk)

DOI: 10.4324/9781003173106

Typeset in Myriad Pro
by codeMantra

Publisher's Note
The publisher accepts responsibility for any inconsistencies that may have arisen during the conversion of this book from journal articles to book chapters, namely the inclusion of journal terminology.

Disclaimer
Every effort has been made to contact copyright holders for their permission to reprint material in this book. The publishers would be grateful to hear from any copyright holder who is not here acknowledged and will undertake to rectify any errors or omissions in future editions of this book.

Contents

Citation Information

The chapters in this book were originally published in the *Journal of European Integration*, volume 42, issue 1 (2020). When citing this material, please use the original page numbering for each article, as follows:

Chapter 1
Renegotiating authority in the Energy Union: A Framework for Analysis
Anna Herranz-Surrallés, Israel Solorio and Jenny Fairbrass
Journal of European Integration, volume 42, issue 1 (2020) pp. 1–17

Chapter 2
Conferring authority in the European Union: citizens' policy priorities for the European Energy Union
Jale Tosun and Mile Mišić
Journal of European Integration, volume 42, issue 1 (2020) pp. 19–38

Chapter 3
EU energy policy integration as embedded intergovernmentalism: the case of Energy Union governance
Pierre Bocquillon and Tomas Maltby
Journal of European Integration, volume 42, issue 1 (2020) pp. 39–57

Chapter 4
Private authority in tackling cross-border issues. The hidden path of integrating European energy markets
Sandra Eckert and Burkard Eberlein
Journal of European Integration, volume 42, issue 1 (2020) pp. 59–75

Chapter 5
Contested energy transition? Europeanization and authority turns in EU renewable energy policy
Israel Solorio and Helge Jörgens
Journal of European Integration, volume 42, issue 1 (2020) pp. 77–93

Chapter 6
Defusing contested authority: EU energy efficiency policymaking
Claire Dupont
Journal of European Integration, volume 42, issue 1 (2020) pp. 95–110

Chapter 7

Power, authority and security: the EU's Russian gas dilemma
Andreas Goldthau and Nick Sitter
Journal of European Integration, volume 42, issue 1 (2020) pp. 111–127

Chapter 8

Gazprom's Nord Stream 2 and diffuse authority in the EU: managing authority challenges regarding Russian gas supplies through the Baltic Sea
Anke Schmidt-Felzmann
Journal of European Integration, volume 42, issue 1 (2020) pp. 129–145

Chapter 9

EU foreign policy and energy strategy: bounded contestation
Richard Youngs
Journal of European Integration, volume 42, issue 1 (2020) pp. 147–162

For any permission-related enquiries please visit:
http://www.tandfonline.com/page/help/permissions

Notes on Contributors

Pierre Bocquillon, School of Politics, Philosophy and Language & Communication Studies, University of East Anglia, Norwich, Norfolk, UK.

Claire Dupont, Department of Public Governance and Management, Ghent University, Belgium.

Burkard Eberlein, Schulich School of Business, York University, Toronto, Canada.

Sandra Eckert, Aarhus Institute of Advanced Studies, Aarhus University, Denmark; Department of Social Sciences, Goethe University, Frankfurt, Germany.

Jenny Fairbrass, Norwich Business School, University of East Anglia, Norwich, Norfolk, UK.

Andreas Goldthau, Willy Brandt School of Public Policy, University of Erfurt, Germany; Institute for Advanced Sustainability Studies, Potsdam, Germany.

Anna Herranz-Surrallés, Faculty of Arts and Social Science, Political Science Department, Maastricht University, the Netherlands.

Helge Jörgens, Department of Political Science and Public Policy, ISCTE - University Institute of Lisbon, Portugal.

Tomas Maltby, Department of Political Economy, King's College London, UK.

Mile Mišić, Institute of Political Science and Heidelberg Centre for the Environment, Heidelberg University, Germany.

Anke Schmidt-Felzmann, Research Centre, General Jonas Žemaitis Military Academy of Lithuania, Vilnius, Lithuania.

Nick Sitter, Law and Governance department, BI Norwegian Business School, Oslo, Norway; School of Public Policy, Central European University, Budapest, Hungary; The Centre for Analysis of Risk and Regulation, London School of Economics, UK.

Israel Solorio, School of Political and Social Sciences, National Autonomous University of Mexico, Mexico City, Mexico.

Jale Tosun, Institute of Political Science and Heidelberg Centre for the Environment, Heidelberg University, Germany.

Richard Youngs, Politics and International Studies, University of Warwick, Coventry, UK.

Renegotiating authority in the Energy Union: A Framework for Analysis

Anna Herranz-Surrallés, Israel Solorio and Jenny Fairbrass (iD)

ABSTRACT
In a context of multiple crises, European Union (EU) energy policy is often identified as one of the few areas still exhibiting strong integration dynamics. However, this policy domain is not exempt from contestation and re-nationalization pressures. This collection seeks to understand better the contradictory integration and fragmentation tendencies by problematizing the notion of *authority*. While authority lies at the heart of European integration theory, less attention has been given to explaining when and why previously conferred authority becomes contested and how authority conflicts are addressed. In framing this collection, we build on sociological approaches to examine systematically the conferral of authority (what counts as authority and how it comes to be recognized) and its contestation (the types of contestation and strategies for managing authority conflicts). We focus this analytical discussion on the Energy Union, being an example of 'hybrid area', which sits uncomfortably at the nexus of different policy areas.

Introduction

In the context of multiple crises that have had centrifugal effects on the European Union (EU), energy policy is often identified as one of the few policy areas that continues to exhibit strong integration dynamics, so much so that it has been labelled as a 'catalyst' for European integration in dangerous times (Delors, Andoura, and Vinois 2015, 1). The *Energy Union* initiative, one of the top priorities of Jean-Claude Juncker's Commission (European Commission 2015) and, more recently, a centre-piece of the *European Green Deal* advanced by Ursula von der Leyen (European Commission 2019), encapsulates this ambition. At the time of its launch in early 2015, the European Commission Vice-president for the Energy Union referred to it as 'undoubtedly the most ambitious European energy project since the European Coal and Steel Community, some 60 years ago' and one that 'has the potential to boost Europe integration the way Coal and Steel did in the 1950s' (Šefčovič 2015). Despite this optimism, however, unlike other recent 'Union' concepts adopted within the EU, such as the Banking or Fiscal Union, the Energy Union has not, so far, led to any additional transfers of competence from the member states to the EU level or the development of new institutions. On the contrary, in some dimensions of EU energy policy the efforts have been in the opposite direction, as member states strive to retain or re-claim authority.

Consequently, EU energy policy seems to capture well the so-called 'post-functionalist' dilemma (Hooghe and Marks 2009). On the one hand, functional efficiency in the provision of public goods, such as financial stability, security or climate change mitigation, requires more governance beyond the state. On the other hand, EU institutions and policies are becoming more politicized and contested domestically. The latest wave of integration theory has explicitly or implicitly attempted to understand better the extent and consequences of this dilemma, by examining the role of crises (Ioannou, Leblond, and Niemann 2015; Schimmelfennig 2017; Tosun, Wetzel, and Zapryanova 2014), politicization (de Wilde, Leupold, and Schmidtke 2016; Costa 2018), or even theorizing (dis)integration (Jones 2018, Vollaard 2014). What brings these approaches together is the conclusion that EU governance is becoming more complex and unpredictable, giving rise to new battle-lines and more hybrid institutional arrangements. So far, it has been unclear whether integration endures against all odds, is receding, or is mutating into new forms.

The special issue introduced here engages with this complexity in a crucial sectoral domain. It does so taking inspiration from global governance theory, which has long tried to understand how societies resolve the tension between the imperative towards cooperation in a globalizing world and the contrary desire to maintain autonomy. Central to those debates is the notion of *authority* beyond the state. Some global governance studies have examined the different ways in which authority is migrating away from states (Kahler and Lake 2004; Rittberger et al. 2008), why authority is conferred, and when it becomes contested (Sending 2015; Zürn 2018). Terms such as 'liquid authority' (Krisch 2017) have recently been coined to capture the growing informal, complex, and unstable relations in global governance.

Our focus is, therefore, on the *renegotiation of authority* in the EU. Anchoring the discussion in global governance theory brings a number of advantages. First, the emphasis on authority allows not only an examination of the formal allocation of competences (often the focus of integration theories) but also of how and why actors gain authority beyond the formal boundaries set by the treaties. Second, it directs our attention to questions about why authority conflicts emerge and how they are managed or mitigated. Finally, it allows us to trace whether contestation leads to actual authority shifts, not only in the vertical direction (upwards or downwards between the local, national and European levels), but also horizontally (between public and private or majoritarian and non-majoritarian actors).

Energy policy is a critical case with which to investigate the transformation of authority patterns in the EU. As a starting point, given that the historical roots of European integration lie in energy cooperation, this policy has a special symbolic weight. Additionally, due to the fact that energy is an area that sits at the cross-roads of different policy domains and areas of competence, ranging from EU exclusive competence (competition policy), to shared competence (climate policy, single market) and intergovernmental domains (security of supply), and includes both an internal and external dimension, it provides a wide range of examples to analyse the extent and consequences of the post-functional dilemma. As a 'flagship initiative' of the Juncker Commission and a crucial pillar for the success of Von der Leyen's European Green Deal, the Energy Union is also a perfect test case for assessing how the EU executive manages this confluence of integrationist and centrifugal pressures.

In framing this collection, this introductory piece aims to accomplish four main tasks. First, it provides an overview of the potential and challenges facing the Energy Union. Second, it develops a novel analytical framework. Third, it summarises the main findings of the volume. Lastly, the article concludes with some forward-looking reflections on EU energy policy and the broader implications for other areas of EU policy-making.

Energy Union: 'saviour' or 'foe' of European integration?

Despite the fact that European integration is rooted in the regional energy cooperation that emerged in the 1950s, for decades, energy was considered as a 'less European' policy area than others (Keay and Buchan 2015, 2). Whilst European energy regulation dates back to the 1970s, it is generally accepted that until recently energy was a 'matter of minor importance on the EU agenda' (Boasson and Wettestad 2013, 1). In fact, the EU did not acquire formal competence concerning energy until the 2009 Treaty of Lisbon, previously secured obliquely via competences associated with competition and environmental policy (Tosun and Solorio 2011).

Over the past ten years, the growing functional necessity for increased cooperation has gradually overcome some of the traditional resistance from national governments in ceding their control over energy issues. Crucially, on the one hand, the energy security crises of late 2000s exposed the vulnerability of individual member states and sparked an EU-wide debate on the need for energy diversification (Herranz-Surrallés 2016). On the other hand, the global demand for urgent action on climate change and the EU's ambition to be an international leader further compelled the need for coordinated action regarding energy among its member states (Wurzel, Connelly, and Liefferink 2017). Moreover, competitiveness pressures made the completion of the internal energy market a priority for the EU (Eikeland 2011). Together these factors facilitated a 'supranational turn' in energy policy, through the 2020 Climate and Energy Package and the Third Internal Energy Market Package adopted in 2009 (Wettestad, Eikeland, and Nilsson 2012, 67). Subsequently, EU institutions have also gradually acquired a central role in securing energy supply, previously a jealously guarded domain of state sovereignty (Maltby 2013). As an example of the growing optimism around EU energy policy, in 2010 Jerzy Buzek and Jacques Delors presented the idea of a European Energy Community, conceived as 'the next chapter in the history of European integration' (Buzek and Delors 2010, 1).

However, despite the hope for a rapid consolidation of a comprehensive and coherent EU energy policy, the initiatives above-mentioned also triggered a debate about the degree of power transferred to the EU, emanating from the member states' reluctance to relinquish their central position with respect to core aspects of the policy (such as the energy mix or relations with external suppliers). The pattern of contested authority claims and counter-claims (i.e. reclamation) among and between member states and EU institutions persists. This became evident during the discussions concerning the 2030 Energy and Climate Framework in 2014, when some member states pressed for less ambitious and less binding targets in comparison to those contained in the 2020 framework, exposing the internal fissures within the EU and between its member states with regard to the policy and its governance (Solorio and Bocquillon 2017, 34–35; Szulecki and

Westphal 2014). The gradual development of an internal EU energy policy also prompted intense political controversy and legal action among external actors, mainly the Russian Federation, which accused the EU of discriminatory actions and of seeking the extra-territorial application of its rules (Kuzemko 2014; Romanova 2016).

With the above-mentioned in mind, the appearance of the concept of 'Energy Union' at the centre stage of the EU's policy agenda was presented as a 'saviour' for the European integration. In April 2014, following the crisis in Ukraine and the Russian intervention in Crimea, the then-Prime Minister of Poland, Donald Tusk, called for the creation of an Energy Union to combat Europe's energy dependence on Russia and return 'the European project to its roots' (Tusk 2014). Much of the emphasis of Tusk's project concerned the security of supply in the gas sector, neglecting the debate about the internal energy market and the climate agenda that had previously characterized the EU's activities in this policy area (Boersma and Goldthau 2017; Szulecki et al. 2016). While Tusk's proposal was not free from criticism, its main merit was to gain media attention and political interest in the notion of the 'Energy Union'.

In this context, the European Commission president, Jean-Claude Juncker, made the Energy Union a top priority on his agenda, widening the range of objectives to include negotiating powers vis-à-vis third countries, as proposed by Tusk, and developing a greater role for renewable energy (Juncker 2014). To coordinate the Commission's efforts, Juncker created the position of vice-president for the Energy Union, a post filled by Maroš Šefčovič. With thirteen new legislative proposals, energy was one of the most dynamic policy areas in the EU during Juncker's administration. In just four years, a new energy governance architecture was re-designed, bundled together by the Regulation on the governance of the Energy Union and Climate Action, adopted in December 2018. Despite the absence of new formal institutions or acts to delegate powers, the development of the Energy Union resulted in the first comprehensive renegotiation of capacities, expectations, and roles in the broad area of energy policy after the formal competence was granted to the EU. It is this renegotiation of authority, undertaken under very difficult conditions, which this collection of papers seeks to disentangle.

Analysing EU governance through the prism of authority

'Authority' is a core concept in Political Science, European Studies and International Relations. In his well-known seminal work to demarcate the domain of Political Science, Eckstein (1973) proposed a characterization of 'politics' as being about *patterns of authority*. Similarly, Schmitter (1970) had previously defined 'regional integration' as the process whereby 'national units come to share part or all of their *decisional authority* with an emerging international organization'. Several decades later, Lake (2010) proposed an understanding of 'global governance' using the prism of *relational authority*. Even though the notion of *authority* has been debated across disciplines over the course of decades, its meaning remains elusive. Crucially, as noted by Krisch (2017, 232) conceptualizations of authority have often been comprehended in restrictive and formalistic ways, as a synonym for the 'ability to make legally binding decisions'. In EU studies, whilst the 1990s' 'governance turn' viewed authority as being dispersed among a variety of levels and actors, it largely treats authority as a legal phenomenon, equivalent to formal competence. For example, Hooghe and Marks' influential definition of the EU as a system of multilevel governance characterised it as a 'layered system of co-existing levels of *authority* – a complex pattern of transnational, public and private institutional relations with overlapping *competences*' (Hooghe and Marks 2003, 235). Similarly, regulatory governance approaches, which focus on agencies and regulatory networks, have usually highlighted formal acts of delegation (Wonka and Rittberger 2010). Studies about 'new' or 'soft' modes of governance have also concentrated on assessing effectiveness in terms of compliance and the relevance of the so-called 'shadow of hierarchy' (Héritier and Lemkuhl 2008). Less attention has been paid to the ways in which different actors acquire authority (be it formally delegated or informally conferred) or how this authority is contested and renegotiated over time.

Significantly, however, greater clarity concerning the concept of authority has been provided by the latest elaborations grounded in sociological approaches to global governance. By contrast with the approaches outlined above, this stream of thought proposes a more dynamic understanding of authority, more broadly defined as an 'ability to induce deference in others' (Krisch 2017, 241; Sending 2015, 21). This more 'liquid' form of authority, to use Krisch's term (2017), encompasses the multiplicity of actors exercising authority in global governance (e.g. private firms, international bureaucrats, non-governmental organizations, professional networks), derived from a range of authority sources beyond mere legal competence. In this context, expertise, capacity, or moral standing may provide actors with the basis for gaining authority beyond the formal delegation of competences. We contend that this perspective provides a promising route to assess current authority debates in the EU for a number of reasons. First, it directs our attention to the fact that authority is in constant flux. For example, Zürn (2018, 8) talks about '*reflexive* authority' to denote that when it comes to governance beyond the state, authority is 'typically not internalized, but it allows a scrutiny of the effects of the exercise of authority at any time'. Similarly, Lake (2010) suggests that the social contract that international authority implies is continuously contested and open to renegotiation, where 'authority is not static, but a dynamic, almost living thing'. Hence, post-national authority always implies some degree of contestation, which must be constantly regained in competition and cooperation with a multiplicity of actors (Sending 2017). In the context of recent severe crises shaking the foundations of the EU, such as the Eurozone, migration, Brexit, and the rising pressure from Euroscepticism on mainstream parties, the assumption that even formally delegated authority is contested gains traction. This is all the more strongly the case in policy areas that remain closely attached to national sovereignty and where EU integration has proceeded in a piecemeal and non-linear fashion, such as energy policy.

Secondly, global governance theory has also emphasised the study of *overlapping* 'spheres of authority' (Rosenau 2007). Contrary to the early focus on 'authority migration' (Gerber and Kollman 2004, 379), which suggests that authority can be relocated from one actor to another, recent approaches argue that authority comes in gradations and with frequent overlaps. On the one hand, overlapping spheres of authority may be seen positively, as necessary and inevitable in solving complex and multi-level problems, as they allow for mutual learning and empowerment of different categories of actors. Regime complexes and hybrid modes of governance are also said to be second-best options for cooperation when power is too dispersed and preferences are too divergent for building robust international regimes (Colgan, Keohane, and Van de Graaf 2012). On the other hand, however, overlying spheres of authority may also spur contestation, particularly when crises occur and issues become politicized. EU energy policy is an example of this type of authority intersection, as it has remained a 'hybrid of co-existing elements', combining strong integration through law and weak integration through coordination (Thaler 2016, 575). The limits of EU and member state authority in this area are, therefore, not clearly fixed and potentially give rise to disagreement and competition.

Finally, sociological approaches to global governance also invite us to examine *informal* types of authority relations, which are not necessarily based on formal acts of delegation. A particularly relevant source of authority in global governance is expertise and competence, known as 'epistemic authority' (Quack 2016; Sending 2015). The energy field is populated by a multiplicity of actors who claim relevant expertise in the governance of the sector. For example, independent regulators at national and European levels play a crucial role in implementing the goals of the Internal Energy Market, as do other industry players such as the Transmission System Operators (TSOs). In addition, the transition to renewable forms of energy is also changing the landscape of authority in the field, in that the possibility that households, co-operatives, and municipalities can produce their own energy is giving rise to new political actors and provoking demands for more decentralized forms of governing (Szulecki 2018).

Based on the considerations above, the following section proposes a novel approach to systematically study authority and its (re)negotiation. The section addresses, in turn, how authority might be *conferred, contested*, and authority conflicts *managed* (see also Table 1).

A framework for analysis: authority conferral, contestation, and conflict management

Conferring authority

Before examining authority relations in a particular period or episode, a useful analytical step is to map the *history of authority patterns*. Following Kahler and Lake (2004, 409), in a multi-level polity such as the EU, this could involve an analysis of the extent to which state authority has been displaced *upwards* to supranational institutions, *downwards* to regions and municipalities, and/or *laterally* to private actors such as companies and NGOs, or non-majoritarian institutions such as independent regulatory agencies. This implies going beyond a discussion of the changes in the distribution of legal competence. Given

Table 1. Summary of the analytical framework.

Dimensions of Authority	Categories of analysis	Measurement
Conferral	*Displacement*	Upwards
		Downwards
		Laterally
	Motivations	Functional needs
		Value-based objectives
Contestation	*Degree*	Low-intensity
		High-intensity
	Type	Sovereignty-based
		Substance-based
Management	*Legal strategies*	Formal adjudication
		Flexibility measures
	Political strategies	(De)politicization
		Enhanced coordination

Source: authors

that authority is a relational concept, authority cannot be considered the property of an actor. As Sending (2015, 5) specifies, a '"source" of authority is not just there for an actor to draw on but must itself be constructed, nurtured, and made effective in particular settings'. Detecting the presence of authority thus requires examining not only authority claims, but also by whom these claims are acknowledged, either through formal recognition or by deference to the rules set and ideas promoted by a given actor.

A second aspect of the analysis is to understand the *causes or reasons* behind the decisions to delegate or defer to certain actors. Both *functional* needs and *value-based* objectives are relevant here. For example, Zürn (2018, 8) defines global governance as 'the exercise of authority across national borders as well as consented norms and rules beyond the state, both of them justified with reference to common goods or transnational problems'. Equally, the displacement of authority away from a member state is often seen as a response to a gap between formal authority and actual capacity to solve an issue. For example, as Hall and Biersteker (2002, 11) argue, when citizens realize that the nation-state can no longer be held accountable on issues that directly affect their lives, 'the exercise of authority by the state is undermined and authority necessarily shifts'. Yet another common expectation is that, in particularly sensitive areas, substantive delegation to international institutions could be demanded by crises, which would oblige states to set aside their sovereignty concerns (Krisch 2017, 245).

Contesting authority

A second dimension of the study is to examine to what extent authority in a given field is contested. Two points of departure exist for this debate. One possibility is that the *degree of contestation* is *low*, either because spheres of authority are clearly delimited, or because even if they are diffuse and overlapping, cooperation and mutual empowerment of different actors operate smoothly and without friction. Where this is the case, this invites reflection on the factors that enable a well-functioning domain despite the assumptions that the exercise of authority beyond the state is likely to generate contestation and competition.

Alternatively, where contestation is *high*, meaning situations where authority conflicts prevent or severely hinder policy making, the next step is to determine the *type of contestation* observed. We define two broad types, which are linked to different dynamics. The first type is *sovereignty-based contestation*, which refers to cases where different actors claim to be the ultimate authority over a particular issue. This type of overlapping authority claim is nicely captured by the notion of 'sovereignty surplus', which denotes common situations in the EU where formal authority is simultaneously claimed by different levels of governance (Walker 2010). The structural cause of this is that the EU can be said to have acquired quasi-sovereign powers, both in a formal and substantive sense: formally, through a gradual process of constitutionalization, setting principles such as the supremacy of EU law and direct effect; and substantively, through its everyday pre-eminence in a wide array of policy sectors (Ibid 2010). This leads to a sovereignty surplus in the sense of 'excess and overlapping quality of claims to sovereignty in the EU (i.e. that ultimate authority is claimed both for the supranational centre and for the member states)' (Walker 2010, 8). Such surpluses are more likely in areas where there is a gap between formal and informal authority, namely issues where EU integration dynamics have extended informally beyond the explicit competences set by treaties, via the exercise of neighbouring or implied competences (Herranz-Surrallés 2014,

958). Yet another common type of sovereignty surplus might manifest as a conflict of jurisdictions between the EU, third countries and/or international law.

The second type is *substance-based contestation*, which derives from how authority is wielded and for what purpose. This contestation does not emerge from competition over legal or decision-making authority as such, but from the erosion of agents' authority when they are perceived as failing to act in accordance with the established social contract or expectations of those having delegated or deferred authority. This contestation pattern might be common in public-private relations: for example, decisions by a government to limit or withdraw authority from regulators or private actors who exercise public func-tions. Another example could be the EU's use of its powers to intervene directly in market or social relations, instead of limiting itself to more regulatory functions, as states have usually expected from the EU (Leuffen, Rittberger, and Schimmelfennig 2013, 5). Episodes revealing incompetence, wrongdoing or discrimination could also lead to the erosion or revocation of authority. Political Science studies have also argued that long periods of depoliticisation via the delegation of functions to private and independent regulatory actors will tend to engender re-politicisation at some point (Flinders and Wood 2015) and give rise to accountability issues (Hall and Biersteker 2002). In sum, unlike sovereignty-based contestation, which denotes competing authority claims on the vertical level (mainly between EU and Member states), substance-based contestation is more likely to imply conflicts of authority on the horizontal dimension, as a result of the ongoing recalibration between public and market actors, or between geopolitical and market-liberal approaches (Goldthau and Sitter 2015; Herranz-Surrallés 2016; Youngs 2011).

Managing authority conflicts

Last but not least, a third dimension of an authority analysis is to examine the strategies employed in mitigating and/or addressing authority contestation and their outcome. We distinguish between legal and political strategies. *Legal strategies* are those aimed at solving authority conflicts by formally (re)allocating actors' authority, and are hence more likely to be used to manage sovereignty-based contestation. The formal recalibration of authority can take place through (a) *formal adjudication*, namely measures that clarify the limits of actors' competences. Within the EU, member states and institutions might bring a case before the Court of Justice of the EU (CJEU) or another form of international arbitration in case of authority conflict with a non-EU country. National Parliaments can also legally prevent the transfer of competence to the EU in the form of subsidiarity checks. Another form of eliminating ambiguity in the allocation of competence is through amending/adopting new legislation or Treaty provisions.

An alternative legal strategy is through (b) *flexibility measures*, having the opposite effect, namely facilitating the dispersion of authority, rather than delimiting it. One such option is through 'micro-differentiation' (de Witte 2017, 25) characterised by high discretion in the implementation of secondary legislation and tailor-made exemptions and derogations (in the energy domain, see Andersen and Sitter 2006; Herranz-Surrallés 2019). Another strategy that facilitates the dispersion of authority is the mixing of governance modes. In that regard, scholars have observed a growing haziness between hard and soft governance (Graziano and Halpern 2016, 5). Energy policy is precisely an area where scholars disagree on whether recent institutional developments in the Energy Union are a sign of 'softening', with Member States' re-gaining control (Solorio and Bocquillon 2017; Thaler 2016) or on the contrary, an example of 'hardening', where the Commission has formally acquired stronger agenda setting and monitoring powers (Ringel and Knodt 2018; Oberthür 2019).

Political strategies are more likely to be employed in substance-based contestation. The very framing of issues can be a powerful tool in the process of re-negotiating authority. The third strategy to manage contestation is therefore (c) *(de)politicisation*. Since substance-based contestation of authority often relates to the balance between public and private authority, one effective strategy to deal with contestation is to seek a recalibration through politicising or depoliticising an issue (Flinders and Buller 2006). Actors favouring greater authority by independent regulatory or market actors will try to *depoliticise* an issue via framing the issue as a technical domain, in order to 'fence off' certain areas of governance from high politics considerations and/or from the involvement of the public/ parliaments. For example, the liberalization of the EU energy markets has been interpreted as a 'de-politicisation strategy' (Eberlein 2010, 65). The opposite strategy is to *(re) politicise* issues. This entails seeking an increase in political control over market or regulatory actors, introducing institutionalized screening procedures, or even the requirement for parliamentary approval in issues such as investment decisions.

Finally, another political strategy that might be used when contestation is mostly substance-based is (d) *enhanced coordination*. Rather than recalibrating powers between public and private actors or between national and EU levels, this strategy would seek to maintain flexible and inclusive arrangements. However, in order to mitigate friction and authority losses, actors may seek to upgrade coordination between the different sites of authority. In EU energy policy, which encompasses a multiplicity of sectors and levels of governance, the existence of coordination structures might be particularly relevant to prevent or mitigate potential authority conflicts.

Reflecting on the contributions of this special issue

The contributors to this special issue have all grappled with the theme of authority in the context of the Energy Union. Together the papers have covered topics as varied as public opinion attitudes towards EU energy policy (Tosun and Misic forthcoming), the new regulation on the governance of the Energy Union and Climate Action (Bocquillon and Maltby forthcoming), the role of private energy transmission operators in the internal market (Eckert and Eberlein forthcoming), the evolution of EU renewable energy and efficiency policies (Solorio and Jorgens forthcoming; Dupont forthcoming), the dilemmas of the EU regulatory power in the gas market (Goldthau and Sitter forthcoming), the local and external dimensions of the Nord Stream 2 pipeline controversy (Schmidt-Felzmann forthcoming), and the impact of foreign policy challenges on EU energy policy (Youngs forthcoming). Despite their different theoretical angles and methodological tools, each paper has successfully highlighted critical patterns relating to the conferral of authority, its contestation and the management of conflict in the Energy Union.

Conferral of authority: growing EU authority beyond formal competence

The overall picture that emerges from this collection is that European energy policy, despite its tardy and peculiar Europeanisation path, has established itself as a central domain of EU activity through a double displacement of authority. On the one hand, several contributions point to an *upward authority shift*, from the member states to the EU. The contributions concerning EU renewable energy and energy efficiency policies (Solorio and Jorgens forthcoming; Dupont forthcoming) illustrate how the EU has, over time, gained significant authority, ahead of the formal recognition of its competence in the Lisbon Treaty. The centrality of the Commission in EU energy policy making is also what leads Bocquillon and Maltby (forthcoming) to argue that 'new intergovernmentalism' inadequately captures the distribution of authority in this sector, proposing instead the notion of 'embedded intergovernmentalism'. Similarly, the EU has also acquired authority in external energy policy (energy security and climate security) through a mixture of exogenous trends and crises as well as a gradual acceptance of the principle that member states should not be able to decide alone about projects that undermine the security of other member states (Goldthau and Sitter forthcoming; Youngs forthcoming). Expertise and the moral high-ground of representing common principles are therefore also relevant sources of EU authority. The consolidation of EU authority is also apparent in the contribution by Tosun and Misic (forthcoming), which indicates that, despite member states' reluctance to transfer energy competences to the EU level, citizens show a very high support for the notion of a common EU energy policy.

On the other hand, two of the contributions highlight the *lateral shift of authority*, from public to private actors, and its interaction with the upward displacement of authority towards the EU. In the internal dimension, Eckert and Eberlein (forthcoming) explore the phenomenon of rising 'private authority' focusing on the operators of the electricity grids, namely the TSOs. The authors find that much of the displacement of authority to these private actors is grounded in their functional expertise and legacy of providing a public good. Again, their involvement in furthering the integration of the European energy market predates by far the formalisation of their role in the EU third electricity directive in 2009. In the external dimension, Goldthau and Sitter (forthcoming) also argue that the EU has acquired a great deal of authority via the exercise of its market-regulatory competences. The authors argue that the Commission's commanding position came precisely from its ability to position itself as a neutral market arbiter as well as its self-restraint in the exercise of regulatory powers. A core idea that the authors advance is therefore that the degree of EU authority depends on whether power is used 'responsibly', namely within the limits of its market-regulatory function.

Contestation of authority: from bounded contestation to sovereignty surpluses

Given the multi-sectoral character of EU energy policy, contributions in this collection find very different levels of contestation, revealing contrasting trends in the context of the Energy Union. On the one hand, the domain of gas supplies is the case that best exemplifies a *high* level of sovereignty-based contestation, which fits well the sovereignty surplus situation, where a multiplicity of actors (local authorities, member states, the EU and third countries) claim decision-making authority over the same issue (Schmidt-Felzmann forthcoming). However, Goldthau and Sitter (forthcoming) also make the case that contestation in gas supplies is not so much about whether the EU has authority to use regulatory policy (sovereignty based contestation) but, rather, the purpose for which the EU should wield its power (substance-based contestation). More specifically, member states appear divided on the question of whether the EU ought to use its regulatory power to address a threat that arises from geopolitics – namely a debate between liberal and geopolitically-oriented approaches. Though less acrimonious, sovereignty-based contestation has also figured prominently in the EU's involvement in the promotion of energy renewables and energy efficiency (Bocquillon and Maltby forthcoming; Dupont forthcoming; Solorio and Jorgens forthcoming), where the limits between the EU's competence for promoting energy sustainability in a well-functioning energy market have often clashed with member states' sovereignty or subsidiarity claims.

On the other hand, some contributions provide examples where sovereignty-based contestation has remained *low* or is decreasing: for example, in the cases of the internal energy market, energy efficiency, and EU external energy policy in general. Concerning the internal energy market, Eckert and Eberlein find that sovereignty-based contestation, for example in network codes and planning, has somewhat decreased alongside the empowerment of private actors, and that substance-based contestation has remained low, despite the suspicion that TSOs might be using their regulatory authority for their own benefit. Dupont (forthcoming) also highlights that the prominent sovereignty-based contestation that marked the early years of EU energy efficiency policy, has gradually subsided since the mid-2000s, and is now more characterised by substance-based contestation, connected to the extent and flexibility of energy efficiency measures. In EU external energy policy, Youngs (forthcoming) also argues that contestation has been lower than was expected when the Energy Union was launched, as many feared then that the EU's more geopolitical focus would intensify tensions. The author advances the idea of 'bounded contestation' to refer to the tempering of differences between institutional actors over external energy strategies.

Management of authority conflicts: towards delimiting or fudging authority?

The papers in this collection have identified a wide range of strategies available to manage or mitigate authority conflicts. The main choice in dealing with authority conflicts seems to be between strategies that aim for a delimitation of authority and strategies that enable its further diffusion. The most direct way of delimiting spheres of authority, *formal adjudication*, reveals a largely unsuccessful strategy to deal with cases of deep-seated contestation. The debate about gas supplies is again the clearest example. As discussed by Schmidt-Felzmann (forthcoming), the Commission struggled for a mandate to negotiate a legal framework with Russia for the construction of the Nord Stream 2 pipeline, yet the legal services of the Council argued against it. Russia also sought international arbitration by bringing cases against the EU internal market rules in several dispute settlement bodies. The Commission's proposal to amend the Third Gas Directive to clarify the application of EU law to sub-sea pipelines entering the EU territory also falls within the formal adjudication category in so far as the aim is to better delimitate the respective spheres of competence between the EU and the member states, and between the EU and third countries/international law. However, to date, these legal interventions have been ineffective in solving the underlying authority conflicts.

A more common strategy to delimit/relocate spheres of authority is politicisation and depoliticisation. *Depoliticisation* efforts have been identified as an effective way to overcome contestation in the integration of the energy market, where framing the role of

TSOs as being purely technical and confined to operational cooperation has contributed to a form of 'hidden integration', effectively sheltered from public attention (Eckert and Eberlein forthcoming). Depoliticisation has also been the strategy of choice for the Commission in addressing the particularly sensitive debate about harmonizing RES support schemes (Solorio and Jorgens forthcoming). On other occasions, the Commission opted for *politicisation* as a method to overcome sovereignty-based contestation and garner support for increasing EU authority. This is the case of energy efficiency, where Dupont (forthcoming) documents the various attempts by the Commission at (re)framing the issue as an 'efficiency-first policy', as part of a long game for solidifying EU competence in this domain. Strategies of (de)politicisation might, therefore, also be a precursor for the use of legal strategies to change the formal distribution of competences between the EU and the member states, or between public and private actors. However, in relation to gas supplies, Goldthau and Sitter (forthcoming) contend that depoliticisation cannot simply be engineered. The authors argue that the standard power-sharing and depoliticisation strategies do not offer viable solutions given that the politicisation of the gas trade is the root of the problem, confronting the EU with a genuine dilemma.

On the other end, several other contributions noted that the management of contestation has often implied greater *flexibility measures* and forms of *enhanced coordination* that, rather than delimit spheres of authority, have facilitated its dispersion. The contributions most closely related to energy sustainability find that the resolution or management of authority conflicts has been achieved via flexibility in implementation and/or new combinations of soft and hard modes of governance (Bocquillon and Maltby forthcoming; Dupont forthcoming; Solorio and Jorgens forthcoming). Yet, they also conclude that the resulting distribution of authority remains unstable and could lead to either further integration or re-nationalisation. Finally, in EU external energy relations, enhanced coordination between energy, foreign policy and climate policy communities is deemed to have been a factor contributing to the mitigation of the high levels of contestation characterising this sensitive domain (Youngs forthcoming).

Conclusion

The new wave of European integration theory, encompassing approaches such as post-functionalism, new intergovernmentalism and new parliamentarism, has revived the discussion about who exercises *power* within the EU and what are the implications for the *legitimacy* of the EU (Schmidt 2018). A focus on authority, broadly understood as ability to induce deference in others, proffers a useful vantage point to these debates. Crucially, rather than espousing a specific 'grand-theoretical' lens, this introductory paper provides a heuristic discussion of the concept of authority, which the articles in this special issue address from a variety of theoretical approaches and methods, including new intergovernmentalism, post-functionalism, regulatory governance, Europeanization theory and/or framing literature. The overall objective of this special issue has been to better diagnose the simultaneous integration and re-nationalisation tendencies in EU energy policy, which due to its multi-sectoral nature, is a focused example of wider patterns of contestation in the EU. In this concluding section, we assess some of the broader theoretical and practical implications of such an approach, both for EU energy policy and for debates in European integration.

The contributions in this collection suggest that the area within the Energy Union that generates the most acrimonious authority conflicts is gas trade. In this domain, the EU faces serious policy dilemmas, which touch on the very *finalité* of European integration. Moreover, attempts to manage contestation by delimiting the respective spheres of competence by formal adjudication have generated further tensions internally and externally. In this context, it is paradoxical that the picture that emerges from this special issue is also a deceleration in EU decarbonisation policies. Particularly when it comes to renewable energy policies, previous Europeanization trends are experiencing negative feedback loops, making some member states more protective of their authority. Notably, the strongest advances of EU authority in this domain, for example the inclusion of renewable support schemes in EU anti-state aid rules, are not particularly helpful in achieving decarbonisation goals.

A second paradox that emerges from this volume is that, while some of the developments in the Energy Union expose the reluctance of member states to cede further authority to the EU, and even reclaim some of it, public opinion seems to mobilise in the opposite direction. Not only are European citizens largely supportive of a common EU energy policy, but also their main priority as regards the Energy Union is the development of renewable energy. In that sense, EU energy policy suffers less from a post-functionalist dilemma, which assumes that the functional need for further integration clashes with a growing resistance from the public, and more from a 'paradox of sovereignty' (McGowan 2009, 21), namely a situation where governments strive to retain their formal authority even though their *de facto* control and capacity to provide public goods is ever more restricted. Echoing Tosun and Misic (forthcoming), there seems to be ample political space for governments to be more ambitious in decarbonisation and for the EU to shift away from a gas-focused external energy policy.

The launch of the European Green Deal framework, which envisages a new binding climate law aimed at achieving carbon neutrality by 2050 (European Commission 2019), seems to capture this sentiment. The implementation of this grand political initiative will soon call for a revival of the debate about the degree of authority that the Commission enjoys and what strategies can best prevent or mitigate the contestation that the ambitious binding 2050 targets are likely to provoke. In this context, future studies could also focus on the impact of other rising sources of authority in EU energy policy not covered in this special issue, such as the recent wave of climate activism that has found particular resonance among the young, as well as the potential for local authorities and prosumer organisations in pushing the boundaries of EU energy policy. In external energy relations, there is also a need to understand better how the EU is equipped for dealing with the international re-allocation of power and authority that energy transitions around the world will bring about.

Beyond EU energy policy, this special issue also contributes to the debate about how to manage authority conflicts in a context where crises and centrifugal tendencies abound. On the one hand, some contributions in this volume demonstrate that formal adjudication is often not a viable strategy to deal with deep-seated authority contestation. On the contrary, it can foster disunity and re-nationalisation pressures. This is, therefore, a call for caution in a context of unprecedented rise in cases brought to the CJEU in delicate areas such as EU external relations (Erlbacher 2017). Depoliticisation strategies, when practicable, can indeed contribute to overcoming authority conflicts. However, they can also lead to policies with a dubious impact on the general interest, as shown by the discussions on 'hidden integration' in the EU energy market or the renewable energy subsidies. On the other hand, the most popular strategies are the ones that foster, rather than limit, the dispersion of authority, such as micro-differentiation and the mixing of governance modes. Yet, these also come with the price tag of fudging political responsibility, as the Energy Union governance regulation exemplifies. Rather than solving this discussion, this special issue represents a move away from a focus on what drives (dis)integration, to debates on what strategies can help manage authority conflicts and their normative and practical consequences.

Acknowledgments

The authors are grateful to the academic association for Contemporary European Studies (UACES) for the funding of the *Collaborative Research Network on European Energy Policy* (2015-2018), which was the breeding ground for this special issue. We are also grateful for *Universiteitsfonds Limburg* (SWOL) for co-funding the authors' workshop in Maastricht, in April 2018. In addition, one of the authors would also like to thank the EU-NormCon (Normative contestation in Europe: Implications for the EU in a changing global order) funded by the Spanish Ministry of Economy and Competitiveness for supporting her participation in the academic events where this paper was presented. Our thanks go also to the several colleagues who provided detailed comments on the papers of this collection: Katja Biedenkopf, Moniek de Jong, Helene Dyrhauge, Rosa Fernández, Quentin Genard, Luca Franza, Rene Kemp, Benjamin Sovacool, Ingmar Versolmann, as well as all the colleagues of the UACES CRN on European Energy Policy. Finally, we are also indebted to an anonymous reviewer and the JEI editors for the helpful comments and guidance on this introductory piece and all of the papers that comprise this special issue.

Disclosure statement

No potential conflict of interest was reported by the authors.

Funding

This work was supported by the University Association of Contemporary European Studies (UACES) [R201486].

ORCID

Jenny Fairbrass http://orcid.org/0000-0001-5292-0720

References

Andersen, S. S., and N. Sitter. 2006. "Differentiated Integration: What Is It and How Much Can the EU Accommodate?" *Journal of European Integration* 28 (4): 313–330. doi:10.1080/07036330600853919.
Boasson, E. L., and J. Wettestad. 2013. *EU Climate Policy: Industry, Policy Interaction and External Environment*. Farnham, UK: Ashgate.
Bocquillon, P., and T. Maltby. forthcoming. "EU Energy Policy Integration as Embedded Intergovernmentalism: The Case of Energy Union Governance Regulation." *Journal of European Integration* 42 (1).

Boersma, T., and A. Goldthau. 2017. "Whiter the EU's Market Making Project in Energy: From Liberalization to Securitization?" In *Energy Union Europe's New Liberal Mercantilism?* edited by S. S. Andersen, A. Goldthau, and N. Sitter, 99–113. New York: Palgrave MacMillan.

Buzek, J., and J. Delors. 2010. *Full Text of the Buzek and Delors Declaration on the Creation of a European Energy Community (EEC)*. Brussels: May 5.

Colgan, J., R. Keohane, and T. Van de Graaf. 2012. "Punctuated Equilibrium in the Energy Regime Complex." *Review of International Organizations* 7 (2): 117–143. doi:10.1007/s11558-011-9130-9.

Costa, O. 2018. "The Politicization of EU External Relations." *Journal of European Public Policy*. early view. doi:10.1080/13501763.2018.1478878).

de Wilde, P., A. Leupold, and H. Schmidtke. 2016. "Introduction: The Differentiated Politicisation of European Governance." *West European Politics* 39 (1): 3–22. doi:10.1080/01402382.2015.1081505.

de Witte, B. 2017. "Variable Geometry and Differentiation as Structural Features of the EU Legal Order." In *Between Flexibility and Disintegration: The Trajectory of Differentiation in EU Law*, edited by B. De Witte, A. Ott, and E. Vos, 9–27. Cheltenham: Edward Elgar Publishing.

Delors, J., S. Andoura, and J. Vinois. 2015. *From the European Energy Community to the Energy Union*. Paris: NOTRE Europe Jacques Delors Institute.

Dupont, C. forthcoming. "Defusing Contested Authority: EU Energy Efficiency Policymaking." *Journal of European Integration* 42 (1).

Eberlein, B. 2010. "Experimentalist Governance in the Energy Sector." In *Experimentalist Governance in the European Union*, edited by C. F. Sabel and J. Zeitlin, 61–78. Oxford: Oxford University Press.

Eckert, S., and B. Eberlein. forthcoming. "Private Authority in Tackling Cross-border Issues. The Hidden Path of Integrating European Energy Markets." *Journal of European Integration* 42 (1). forthcoming.

Eckstein, H. 1973. "Authority Patterns: A Structural Basis for Political Inquiry." *American Political Science Review* 67 (4): 1142–1161. doi:10.2307/1956537.

Eikeland, P. O. 2011. "The Third Internal Energy Market Package: New Power Relations among Member States, EU Institutions and Non-state Actors?" *Journal of Common Market Studies* 49 (2): 243–263. doi:10.1111/j.1468-5965.2010.02140.x.

Erlbacher, F. 2017. "Recent Case Law on External Competences of the European Union: How Member States Can Embrace Their Own Treaty." *CLEER 2017/2*. https://www.asser.nl/media/3485/cleer17-2_web.pdf

European Commission. 2015. "A Framework Strategy for A Resilient Energy Union with A Forward-Looking Climate Change Policy." *COM/2015/080 final*, February 25

European Commission. 2019. "The European Green Deal." *COM(2019) 640 final*, December 11.

Flinders, M., and J. Buller. 2006. "Depoliticization, Democracy and Arena Shifting." In *Autonomy and Regulation. Coping with Agencies in the Modern State*, edited by T. Christensen and P. Laergeid, 53–80. Cheltenham: Edward Elgar.

Flinders, M., and M. Wood. 2015. "When Politics Fails: Hyper-Democracy and Hyper-Depoliticization." *New Political Science* 37 (3): 363–381. doi:10.1080/07393148.2015.1056431.

Gerber, E. R., and K. Kollman. 2004. "Introduction – Authority Migration: Defining an Emerging Research Agenda." *Political Science and Politics* 37 (3): 397–401.

Goldthau, A., and N. Sitter. 2015. "Soft Power with a Hard Edge: EU Policy Tools and Energy Security." *Review of International Political Economy* 22 (5): 941–965. doi:10.1080/09692290.2015.1008547.

Goldthau, A., and N. Sitter. forthcoming. "Power, Authority and Security: The EU's Russian Gas Dilemma." *Journal of European Integration* 42 (1). forthcoming.

Graziano, P. R., and C. Halpern. 2016. "EU Governance in Times of Crisis: Inclusiveness and Effectiveness beyond the 'hard' and 'soft' Law Divide." *Comparative European Politics* 14 (1): 1–19. doi:10.1057/cep.2015.6.

Hall, R. B., and T. J. Biersteker. 2002. "The Emergence of Private Authority in the International System." In *The Emergence of Private Authority in Global Governance*, edited by R. B. Hall and T. J. Biersteker, 3–22. Cambridge: Cambridge University Press.

Héritier, A., and D. Lemkuhl. 2008. "Introduction: The Shadow of Hierarchy and New Modes of Governance." *Journal of Public Policy* 28 (1): 1–17. doi:10.1017/S0143814X08000755.

Herranz-Surrallés, A. 2014. "The EU's Multilevel Parliamentary (Battle)field: Inter-parliamentary Cooperation and Conflict in the Area of Foreign and Security Policy." *West European Politics* 37 (5): 957–975. doi:10.1080/01402382.2014.884755.

Herranz-Surrallés, A. 2016. "An Emerging EU Energy Diplomacy? Discursive Shifts, Enduring Practices." *Journal of European Public Policy* 23 (9): 1386–1405. doi:10.1080/13501763.2015.1083044.

Herranz-Surrallés, A. 2019. "Energy Policy and European Union Politics." *Oxford Research Encyclopedia of Politics*. doi:10.1093/acrefore/9780190228637.013.1079.

Hooghe, L., and G. Marks. 2003. "Unraveling the Central State, but How? Types of Multi-level Governance." *American Political Science Review* 97 (2): 233–243.

Hooghe, L., and G. Marks. 2009. "A Postfunctionalist Theory of European Integration: From Permissive Consensus to Constraining Dissensus." *British Journal of Political Science* 39 (1): 1–23. doi:10.1017/S0007123408000409.

Ioannou, D., P. Leblond, and A. Niemann. 2015. "European Integration and the Crisis: Practice and Theory." *Journal of European Public Policy* 22 (2): 155–176. doi:10.1080/13501763.2014.994979.

Jones, E. R. 2018. "Towards a Theory of Disintegration." *Journal of European Public Policy* 25 (3): 440–451. doi:10.1080/13501763.2017.1411381.

Juncker, J. C. 2014 "A New Start for Europe. Opening Statement in the European Parliament Plenary Session." *Strasbourg*, July 15.

Kahler, M., and D. A. Lake. 2004. "Governance in a Global Economy: Political Authority in Transition." *Political Science and Politics* 37 (3): 409–414.

Keay, M., and D. Buchan. 2015. "Europe's Energy Union – A Problem of Governance."*Oxford Energy Comment*. Oxford: Oxford Institute for Energy Studies.

Krisch, N. 2017. "Liquid Authority in Global Governance." *International Theory* 9 (2): 237–260. doi:10.1017/S1752971916000269.

Kuzemko, C. 2014. "Ideas, Power and Change: Explaining EU–Russia Energy Relations." *Journal of European Public Policy* 21 (1): 58–75. doi:10.1080/13501763.2013.835062.

Lake, D. A. 2010. "Rightful Rules: Authority, Order, and the Foundations of Global Governance." *International Studies Quarterly* 54 (3): 587–613. doi:10.1111/isqu.2010.54.issue-3.

Leuffen, D., B. Rittberger, and F. Schimmelfennig. 2013. *Differentiated Integration. Explaining Variation in the European Union*. Basingstoke: Palgrave.

Maltby, T. 2013. "European Union Energy Policy Integration: A Case of European Commission Policy Entrepreneurship and Increasing Supranationalism." *Energy Policy* 55: 435–444. doi:10.1016/j.enpol.2012.12.031.

McGowan, F. 2009. "International Regimes for Energy: Finding the Right Level for Policy." In *Energy for the Future*, edited by I. Scrase and G. Mackerron, 20–34. New York: Palgrave, Macmillan.

Oberthür, S. 2019. "Hard or Soft Governance? the EU's Climate and Energy Policy Framework for 2030." *Politics and Governance* 7 (1): 17–27. doi:10.17645/pag.v7i1.1796.

Quack, S. 2016. "Expertise and Authority in Transnational Governance." In *Authority in Transnational Legal Theory. Theorising across Disciplines*, edited by R. Cotterrell and M. Del Mar, 361–386. Cheltenham: Edward Elgar.

Ringel, M., and M. Knodt. 2018. "The Governance of the European Energy Union: Efficiency, Effectiveness and Acceptance of the Winter Package 2016." *Energy Policy* 112: 209–220. doi:10.1016/j.enpol.2017.09.047.

Rittberger, V., M. Nettesheim, C. Huckel, and T. Göbel. 2008. "Introduction: Changing Patterns of Authority." In *Authority in the Global Political Economy*, edited by V. Rittberger and M. Nettesheim, 1–9. Basingstoke: Palgrave.

Romanova, T. 2016. "Is Russian Energy Policy Towards the EU Only about Geopolitics? the Case of the Third Liberalisation Package." *Geopolitics* 21 (4): 857–879. doi:10.1080/14650045.2016.1155049.

Rosenau, J. N. 2007. "Governing the Ungovernable: The Challenge of a Global Disaggregation of Authority." *Regulation & Governance* 1: 88–97. doi:10.1111/rego.2007.1.issue-1.

Schimmelfennig, F. 2017. "Theorising Crisis in European Integration." In *The European Union in Crisis*, edited by D. Dinan, N. Nugent, and W. E. Paterson, 316–336. Basingstoke: Palgrave Macmillan.

Schmidt, V. 2018. "Rethinking EU Governance: From 'old' to 'new' Approaches to Who Steers Integration." *Journal of Common Market Studies* 56 (7): 1544–1561. doi:10.1111/jcms.12783.

Schmidt-Felzmann, A. forthcoming. "Nord Stream 2 and Diffuse Authority in the EU: Managing Authority Challenges regarding Russian Pipelines in the Baltic Sea Area." *Journal of European Integration* 42 (1).

Schmitter, P. C. 1970. "A Revised Theory of Regional Integration." *International Organization* 24 (4): 836–868. doi:10.1017/S0020818300017549.

Šefčovič, M. 2015. "Driving the EU Forward: The Energy Union." *SPEECH-15-4520_EN*, February 27.

Sending, O. J. 2015. *The Politics of Expertise. Competing for Authority in Global Governance.* Ann Arbor: Univeristy of Michigan Press.

Sending, O. J. 2017. "Recognition and Liquid Authority." *International Theory* 9 (2): 311–328. doi:10.1017/S1752971916000282.

Solorio, I., and P. Bocquillon. 2017. "EU Renewable Energy Policy: A Brief Overview of Its History and Evolution." In *A Guide to EU Renewable Energy Policy. Comparing Europeanization and Domestic Policy Change in EU Member States*, edited by I. Solorio and H. Jorgens, 23–42. Cheltenham: Edward Elgar.

Solorio, I., and H. Jorgens. forthcoming. "Contested Energy Transition? Europeanization and Authority Turns in EU Renewable Energy Policy." *Journal of European Integration* 42 (1).

Szulecki, K. 2018. "Conceptualizing Energy Democracy." *Environmental Politics* 27 (1): 21–41. doi:10.1080/09644016.2017.1387294.

Szulecki, K., and K. Westphal. 2014. "The Cardinal Sins of European Energy Policy: Non-Governance in an Uncertain Global Landscape." *Global Policy* 5 (1): 38–51. doi:10.1111/gpol.2014.5.issue-s1.

Szulecki, K., S. Fischer, A. T. Gullberg, and O. Sartor. 2016. "Shaping the 'energy Union': Between National Positions and Governance Innovation in EU Energy and Climate Policy." *Climate Policy* 16 (5): 548–567. doi:10.1080/14693062.2015.1135100.

Thaler, P. 2016. "The European Commission and the European Council: Coordinated Agenda Setting in European Energy Policy." *Journal of European Integration* 38 (5): 571–585. doi:10.1080/07036337.2016.1178252.

Tosun, J., A. Wetzel, and G. Zapryanova. 2014. "The EU in Crisis: Advancing the Debate." *Journal of European Integration* 36 (3): 195–211. doi:10.1080/07036337.2014.886401.

Tosun, J., and I. Solorio. 2011. "Exploring the Energy–environment Relationship in the EU: Perspectives and Challenges for Theorizing and Empirical Analysis." *European Integration Online Papers* 15 (1). http://eiop.or.at/eiop/pdf/2011-007.pdf.

Tosun, J., and M. Misic. forthcoming. "Conferring Authority in the European Union: Citizens' Policy Priorities for the European Energy Union." *Journal of European Integration* 42 (1).

Tusk, D. 2014. "A United Europe Can End Russia's Energy Stranglehold." *Financial Times*, April 21.

Vollaard, H. 2014. "Explaining European Disintegration." *Journal of Common Market Studies* 52 (5): 1142–1159. doi:10.1111/jcms.12132.

Walker, N. 2010. "Surface and Depth: The EU's Resilient Sovereignty Question." In *Political Theory of the European Union*, edited by J. Neyer and A. Wiener, 91–110. Oxford: Oxford University Press.

Wettestad, J., P. O. Eikeland, and M. Nilsson. 2012. "EU Climate and Energy Policy: A Hesitant Supranational Turn." *Global Environmental Politics* 12 (2): 67–86. doi:10.1162/GLEP_a_00109.

Wonka, A., and B. Rittberger. 2010. "Credibility, Complexity and Uncertainty: Explaining the Institutional Independence of 29 EU Agencies." *West European Politics* 33 (4): 730–752. doi:10.1080/01402381003794597.

Wurzel, R., J. Connelly, and D. Liefferink, eds. 2017. *The European Union in International Climate Change Politics. Still Taking A Lead?* London: Routledge.

Youngs, R. 2011. "Foreign Policy and Energy Security: Markets, Pipelines and Politics." In *Toward a Common European Union Energy Policy*, edited by J. S. Duffield and V. L. Birchfield, 41–60. New York: Palgrave Macmillan.

Youngs, R. forthcoming. "EU Foreign Policy and Energy Strategy: Bounded Contestation." *Journal of European Integration* 42 (1).

Zürn, M. 2018. *A Theory of Global Governance. Authority, Legitimacy, and Contestation.* Oxford: Oxford University Press.

Conferring authority in the European Union: citizens' policy priorities for the European Energy Union

Jale Tosun and Mile Mišić (iD)

ABSTRACT
We analyse data from the 2018 Eurobarometer survey to provide a debate on EU authority in the field of energy policy beyond the member states' preferences. More specifically, we look at why citizens are willing to confer authority to the EU in making energy policy and which policy priorities they indicate. Focusing on public opinion appears promising as the data reveals that European citizens have a more positive stance on allocating policy competences to the EU than the member states' governments. The descriptive analysis shows that most citizens prefer the Energy Union to prioritise the promotion of renewable energy. This holds particularly true for citizens living in Western Europe that have a left-leaning ideology and who perceive climate change as an issue. For Central-East European citizens, especially those who are right-leaning and perceive energy security as a problem, the Energy Union should give priority to increasing energy security.

1. Introduction

With the enactment of the Treaty of Lisbon in December 2009, EU institutions, among these most notably the European Commission, became equipped with a comprehensive mandate for proposing and passing energy policies (Tosun and Solorio 2011; Solorio and Morata 2012; Wettestad, Eikeland, and Nilsson 2012; Maltby 2013; Boasson and Wettestad 2013; Kustova 2017). For many years, the member states were hesitant to delegate their policymaking competences, especially in matters related to energy security (Pointvogl 2009; Eberlein 2012; Szulecki et al. 2016). However, in 2014 the incoming President of the European Commission, Jean-Claude Juncker, proposed 'a new European Energy Union' (Szulecki et al. 2016, 552), building on an idea launched by then Prime Minister of Poland, Donald Tusk (see, Austvik 2016). In February 2015, the Commission presented its 'Framework Strategy for the Energy Union', which comprises the following main dimensions (European Commission 2015, 4): energy security, solidarity and trust; a fully integrated energy market; energy efficiency that contributes to the modernisation of demand; decarbonisation of the EU's economy; and research, innovation and competitiveness.

Despite the recent push towards further integration in the realm of energy policy (Szulecki et al. 2016, 553), the member states 'continue to determine the conditions for exploiting their energy resources, their choice among different energy sources and the general structure of their energy supply' (Ringel and Knodt 2018, 210), as enshrined in the Lisbon Treaty (194.2 TFEU). In fact, as the introduction to this volume (Herranz-Surrallés, Solorio, and Fairbrass forthcoming) alludes to, in some fields of energy policy (e.g. renewable energy) the member states have strived to regain control (see also Bocquillon and Maltby forthcoming; Solorio and Jörgens forthcoming). Put differently, some member states have aimed to renegotiate the policymaking authority of the EU in the field of energy policy (Herranz-Surrallés, Solorio, and Fairbrass forthcoming).

Remarkably, despite the member states' reluctance to transfer further policymaking competences to the EU, European citizens appear rather receptive to the notion of a common energy policy. Indeed, almost three-quarters (73%) of Europeans surveyed in 2018 for the 89.1 Eurobarometer indicated that they support a common energy policy among the EU member states (European Commission 2018a, 8). While one could argue that this degree of public support could be an outlier and therefore be limited to the 2018 data only, the Eurobarometer report shows that the level of support for a common energy policy has never dipped below 70% since 2014 (European Commission 2018a, 8). We find this consistently high degree of public support for a common energy policy puzzling and consider it worth investigating.

Which policy objectives do European citizens associate with the Energy Union? Which particular objective dominates their views on it? The first two questions reflect the multidimensional character of the Energy Union as outlined above. In this analysis, however, we concentrate on the dimensions related to energy security and decarbonisation. Our choice for these two dimensions is an analytical one since existing empirical research shows that climate change and energy security are perceived as separate problems in most countries, even though they are interlinked (see Toke and Vezirgiannidou 2013; Cherp and Jewell 2014).

Renewable energy plays an important role for both energy security and decarbonisation, but its production can potentially harm the environment (e.g. through land-use effects) (European Commission 2015, 16; see also Tosun and Solorio 2011). Therefore, when concentrating on energy security and decarbonisation, it is also important to pay attention to environmental concerns, which is reflected in the third research question: Which factors explain whether citizens expect the European Energy Union to give priority to increasing energy security, fighting climate change, or protecting the environment?

Our empirical investigation relies on data from the 89.1 Eurobarometer survey (European Commission 2018b) and shows that most EU citizens believe the Energy Union's top priority should be the development of renewable energy – an area, remarkably, in which the EU's policymaking competence is being increasingly contested (see Bocquillon and Maltby forthcoming; Solorio and Jörgens forthcoming). Environmental protection and the fight against global warming rank as the second and third priorities respectively. Fewer citizens deem the Energy Union's top priority to lie in securing energy supply and ensuring the EU's energy independence. In terms of the explanatory factors, the image of the EU and an understanding of how the EU functions, together with problem perception, are the most robust factors for explaining the citizens' preferences regarding which targets they believe should be prioritised by the European Energy Union.

The citizens' preferences also vary between the West European and Central-East European member states.

The remainder of this study unfolds as follows. First, we show how political attention to energy issues at the EU level has developed over time. That exercise will help us to better situate this study of public opinion in the context of research that concentrates on how citizens perceive energy policy to be made in the EU. Next, we present the theoretical framework on which our analysis rests, which is followed by some clarifications on the measurement of our key concepts. We then discuss the findings obtained by estimating multi-level logit models. Finally, we summarise our main findings and offer some concluding remarks.

2. Political attention to energy issues: substantiating the empirical puzzle

This section serves to underscore the puzzle alluded to in the introduction, namely that EU citizens consistently confer high authority on the EU when it comes to energy policy, despite the ups and downs in political attention and the general reluctance of member states to delegate further powers. While Herranz-Surrallés, Solorio, and Fairbrass (forthcoming) addresses the latter point, we concentrate on political attention in this section. To this end, we rely on a dataset prepared by Alexandrova et al. (2014), which contains information on the number of mentions of policy issues in the conclusions of meetings held by the European Council, which wields an important, informal agenda-setting power (Bocquillon and Dobbels 2014). It is important to note that policy issues discussed at European Council meetings do not necessarily lead to the formation of legislative initiatives by the EU Commission, but they are on the discussion agenda and are subject to serious consideration and deliberation.

We present the data as provided by Alexandrova et al. (2014) and have added observations for the years 2015–2017 by employing the same methodology as laid out in the corresponding codebook. The codebook has 23 major topics, which each contain various subtopics. We broke down the texts into single units of analysis (quasi-sentences) and categorised each of these according to the codebook. To capture energy-related content, we used topic 8 (Energy), which has nine subtopics, including fossil fuels, renewables, energy efficiency, and nuclear energy.

Figure 1 reveals the variation in political attention to energy issues over time. While energy issues received enhanced attention in the late 1970s, over the 1980s there were very few mentions of energy issues in the conclusions of the European Council meetings. During the 1990s and 2000s, there were some peaks in the political attention paid to energy issues. The picture changes noticeably after the year 2006 for energy issues, as indicated by the peaks and generally elevated levels of attention. The number of mentions for energy peaked in 2006 and then gradually decreased, before peaking again in 2011 and particularly in 2014.

The peak in 2006 stems from the heads of government calling on the European Commission at the Spring European Council to develop – on the basis of its Green Paper 'European Strategy for Sustainable, Competitive and Secure Energy' – an 'Energy Policy for Europe', which many observers consider the beginning of EU energy policy (see, e.g. Solorio 2011). The European Council (2006, 13–17) invited the Commission to establish a consistent set of strategies for the EU's internal and external energy policy (see

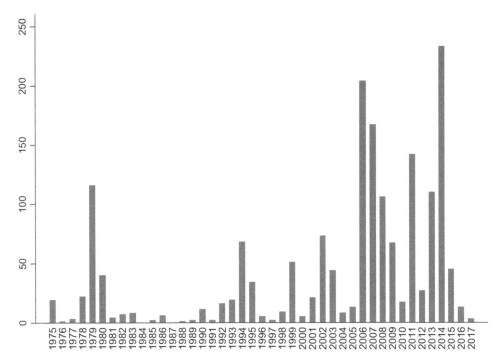

Figure 1. Political attention to energy policy.
Notes: Own elaboration based on Alexandrova et al. (2014) and own research for the years 2015–2017.

Herranz-Surrallés 2016, 2017) in order to increase the security of energy supply, competitiveness and environmental sustainability. The nuclear accident at the *Fukushima Daiichi* nuclear power plant in 2011 resulted in the mentions of energy issues increasing as the heads of government deliberated on the implementation of stress tests for nuclear power plants and on higher standards for nuclear safety (see Álvarez-Verdugo 2015).

Three factors explain the peak in the mentions of energy issues in the European Council conclusions in 2014. First, the negotiation of the 2030 energy and climate framework (see Szulecki 2016); second, the coordination of the EU's negotiation position at the Conference of the Parties of the United Nations Framework Convention on Climate Change to be held in 2015 in Paris (see Biedenkopf and Petri 2019); third, deliberations on the EU's positions on the energy-related implications of the Russia-Ukraine conflict over Crimea (see Szulecki and Westphal 2018).

There are several messages to be taken from this figure: first, energy policy has been the subject of political debate among the member states since the 1970s. Second, political attention to energy issues increased sharply from 2006 onwards. The timing of the increase is plausible, for ten new member states joined the EU in 2004, all of which were highly dependent on energy supplies, meaning the issue of energy security was brought onto the political agenda (see Mišík 2015). In addition, the formal beginning of EU energy policy in 2007 and the integration of energy policy with climate policy also resulted in enhanced political attention (see Solorio and Morata 2012).

Yet, since 2014, energy issues have received notably less attention from the European Council, despite the proposal for the creation of the Energy Union and the 'Winter

Package', which addresses the governance issues related to the Energy Union (see Ringel and Knodt 2018). One of the reasons for this decline could be the growing hesitation of some member states to transfer further competences for making energy policy to the EU (Herranz-Surrallés, Solorio, and Fairbrass forthcoming). This development at the political level makes the high level of public support for a common energy policy even more puzzling.

3. Theoretical framework and hypotheses

The subsequent empirical analyses address two complementary aspects: first, they provide a descriptive overview of the citizens' opinions on policy priorities for the Energy Union; second, we test which factors explain the Europeans' differing views on the policy priorities of the Energy Union.

This theory section prepares the second analysis, which focuses on individual- and country-level variables that could explain the differences in the respondents' opinions regarding the policy priorities of the Energy Union. As explained in the introduction, the Energy Union consists of multiple dimensions (see also Tosun, Zöckler, and Rilling 2019). When inspecting the framework strategy for the Energy Union, it becomes apparent that the individual dimensions consist of many sub-dimensions (see European Commission 2015). However, in order to be able to offer a theoretically informed empirical analysis and to formulate testable hypotheses, we must limit the scope of this analysis. The relationship between energy security and climate change appears particularly promising for that purpose (e.g. Toke and Vezirgiannidou 2013).

The inclusion of considerations related to the affordability of energy, which is also covered in the Eurobarometer survey, would have entailed a different theoretical approach. For example, it would have necessitated engagement with the concept of 'energy poverty' (see Bouzarovski and Tirado Herrero 2017) and the inclusion of different explanatory variables (e.g. spatial distribution of energy poverty). Hence the development of a theoretical framework that includes energy affordability would have gone well beyond the scope of this analysis and the special issue to which it contributes. The same goes for the other dimensions of the Energy Union as put forth by the European Commission (2015). However, when concentrating on energy security, climate change and environmental protection, we can develop a coherent theoretical model.

We will examine five outcome variables, which we assign to three overarching categories in order to facilitate a more straightforward presentation and discussion of the policies that the citizens believe to be priorities of the Energy Union:

- Security dimension: guaranteeing a continuous supply of energy; guaranteeing the EU's independence in the field of energy
- Climate dimension: developing renewable energy; fighting global warming
- Environment dimension: protecting the environment

We theorise that two individual-level determinants and two country-level determinants are important for explaining these outcome variables. The explanatory variables at the individual level concern ideological orientation and problem perception. The second set of explanatory variables refers to domestic problem-solving performance and a differentiation between West European and Central-East European member states. We will now discuss each of these variables in turn and put forth the corresponding hypotheses.

Individual-level variables

We regard the public's views on the Energy Union to be determined by how they perceive of problems, but also by how the problem perception is potentially influenced by ideological predispositions. Therefore, the first potential factor for explaining differences in the individuals' policy priorities for the Energy Union refers to political ideology. As Neumayer (2004) argues, a pro-environmental orientation aligns with distributional concerns and demand for state intervention in the economy, which is the dimension that has been widely regarded – and shown – to separate the political left from the political right. Consequently, we expect individuals with a left-wing political orientation to differ from individuals with right-wing orientation in the sense that the latter prefer the Energy Union to give less priority to the goal of environmental protection. Turning to the climate-related dimension of the Energy Union, we also expect individuals with a right-wing political orientation to be less supportive of its aims to develop renewable energy and fight climate change than individuals with a left-wing orientation. One reason for this expectation resonates with the explanation given by Neumayer (2004) for pro-environmental attitudes: protecting the climate requires government regulation of the economy. Clements (2012) alludes to another mechanism underlying this relationship, namely that right-wing individuals are more sceptical of climate change and are therefore likely to give less priority to climate protection.

While the impact of political ideology on pro-environmental and pro-climate attitudes has been researched extensively, the impact of political ideology on energy security has received scant attention only. To be able to formulate a hypothesis, it is useful to take into consideration the study by Latré, Thijssen, and Perko (2019), which shows that right-wing parties support nuclear power plants as a strategy for increasing the security of energy supply. This induces us to expect right-wing individuals to give priority to energy security over environmental and climate protection. More generally, security issues are associated with the right-wing ideology (see Budge 2013). Consequently, we expect individuals with a right-wing orientation to be more likely to give priority to the security-related dimension of the Energy Union.

H1a: Individuals with a right-wing orientation are more likely to give priority to the security-related dimension of the Energy Union.

H1b: Individuals with a right-wing orientation are less likely to give priority to the climate-related dimension of the Energy Union.

H1c: Individuals with a right-wing orientation are less likely to give priority to the environment-related dimension of the Energy Union.

The second explanatory variable refers to the individual's problem perception and aligns directly with the country-level variable on domestic problem-solving performance. Our reasoning concerning this variable, and the one for the country level, is derived from policy feedback theory (see Mettler and Soss 2004), which has been explored empirically in the comparative study of attitudes to the welfare state (see, e.g. Kumlin and Stadelmann-Steffen 2014). This literature argues that citizens' attitudes can be the result of national policies. We modify this argument and hypothesise that absent or insufficient policies to address certain issues are likely to affect how these issues are perceived. If issues are perceived as problematic (due to lacking or insufficient policy responses), they should affect the citizens' opinions on future policies. Therefore, individuals who consider energy security, climate change or environmental degradation a problem should give priority to these issues.

H2a: Individuals who consider energy dependence an important issue are likely to give priority to the security-related dimension of the Energy Union.

H2b: Individuals who consider climate change an important issue are likely to give priority to the climate-related dimension of the Energy Union.

H2c: Individuals who consider environmental degradation an important issue are likely to give priority to the environment-related dimension of the Energy Union.

Country-level variables

Research on public opinion has emphasised that attitudes to issues do not only vary within countries but also across them (e.g. Braun and Tausendpfund 2014). One of the reasons for cross-country differences in public opinion can be the variation in perception of the domestic government's problem-solving performances, which aligns with the reasoning of the policy feedback theory outlined above (see Mettler and Soss 2004; Kumlin and Stadelmann-Steffen 2014; Stadelmann-Steffen and Eder 2020). Respondents in countries with a high degree of energy dependence should give priority to increasing the security of energy supply. Likewise, respondents in countries with a poor environmental policy performance should prioritise the environment-related dimension of energy policy, and respondents in countries with a poor climate policy performance the climate-related dimension of the Energy Union.

To derive our hypotheses, we modify the policy feedback theory to differentiate between public demand for domestic and EU-wide policy measures. If the problem-solving performance of the national government is considered weak and becomes observable through high degradation or a high level of energy dependence, citizens could become receptive to the idea of allocating the respective policy competence to the EU level (see, e.g. Maggetti and Trein 2019).

H3a: The worse the national policy performance in energy security, the more likely citizens are to give priority to the security-related dimension of the Energy Union.

H3b: The worse the national climate policy performance, the more likely citizens are to give priority to the climate-related dimension of the Energy Union.

H3c: The worse the national environmental policy performance, the more likely citizens are to give priority to the environment-related dimension of the Energy Union.

The literature on EU energy policy has revealed differences in citizens' opinions between country groups, such as West European and Central-East European member states. Adapting this reasoning to the case at hand, we expect citizens in West European member states to be more supportive of strengthening the environment- and climate-related aspects of the Energy Union (see McCright, Dunlap, and Marquart-Pyatt 2016), whereas the respondents in the Central-East European member states should in theory be more willing to give priority to increasing the security of energy supply. Concerning the latter, Mišík (2015) showed that even policymakers in Central-East Europe are willing to confer authority on the EU if this results in greater energy security. Therefore, when choosing between different energy policy priorities, it appears plausible to expect an East-West divide among the respondents.

H4a: Citizens in Central-East Europe are more likely to give priority to the security-related dimension of the Energy Union.

H4b: Citizens in Central-East Europe are less likely to give priority to the climate-related dimension of the Energy Union.

H4c: Citizens in Central-East Europe are less likely to give priority to the environment-related dimension of the Energy Union.

4. Measurement of the variables and methodology

To address our research questions, we rely on survey data for the EU member states. The primary data source is the Eurobarometer 89.1 standard survey fielded in 2018. All interviews for the 89.1 survey were conducted in March 2018 in the national language of the respective country, using face-to-face interviewing techniques. The number of respondents in each country, except Malta, Cyprus and Luxembourg, was at least 1,000.

Our five outcome variables concern the indication of priority given to Energy Security, Energy Independence, Climate Protection, Renewable Energy, and Environmental Protection. The first two variables on energy security and independence refer to the security dimension (see also European Commission 2015, 4). Climate protection and renewable energy constitute the climate-dimension. For the environment dimension, we can rely on one indicator only since the other answer options in the pertinent Eurobarometer question incorporate other dimensions of the Energy Union (see Table A1 in the Appendix). We operationalise all outcome variables by relying on question QB2 in the survey, which reads as follows: *In your opinion, which of the following objectives should be given top priority in a European energy union?* The variables indicate whether the respondents chose the respective response category (coded as 1) or not (coded as 0).

The focal explanatory variables at the individual level are Ideology and Problem Perception. The following question gauges the variable Ideology: *In political matters, people talk of 'the left' and 'the right'. How would you place your views on this scale?*

(Question D1). The scale is a 10-point left-right scale which runs from 1 (left) to 10 (right), thereby corresponding to a standard measurement in the literature (see, e.g. Clements 2012).

For Problem Perception, we assess three dimensions that refer to environmental pollution, climate change, and energy security. They are based on the respondents' answers to the following question: *What do you think are the two most important issues facing the EU at the moment?* (QA5). The respondents could choose two out of 16 answer categories. Our three binary variables are based on answer categories: (11) 'The environment', (12) 'Energy supply', and (13) 'Climate change' (see Table A1 in the Appendix).

In accordance with the pertinent literature on public opinion on climate change (see, e. g. Stadelmann-Steffen and Eder 2020) and the environment (see, e.g. Neumayer 2004), we include a whole battery of control variables, such as Age and Gender, for which the measurement should be straightforward. We also add a number of EU-related control variables: EU Image, EU Democracy, Country Democracy, and EU Knowledge. These are variables we consider important with regard to the energy policy priorities expressed by citizens, but for which we cannot formulate hypotheses that would differentiate between the different policy priorities expressed by the respondents. These variables assess the respondent's general opinion and knowledge on the functioning of the EU as well as their (dis)satisfaction with democracy in the EU and in their respective member state. We include satisfaction with democracy in one's country and with the EU since we might be able to observe a 'venue shopping' effect. Citizens satisfied with how democracy works in their own countries could be generally less inclined to allocate policymaking competences to the EU, whereas citizens who are satisfied with democracy in the EU could be inclined to allocate more policymaking competences to that level (see, e.g. Hadjar and Beck 2010).

EU Image is based on the following question: *In general, does the EU conjure up for you a very positive, fairly positive, neutral, fairly negative or very negative image?* (Question QA9). Each answer is assigned a numerical value and the (reverted) scale goes from 1 (very negative) to 5 (very positive).

Country Democracy is measured by means of the following question: *On the whole, are you very satisfied, fairly satisfied, not very satisfied or not at all satisfied with the way democracy works in (your country)?* (Question QA17a). EU Democracy is measured by means of this question: *And how about the way democracy works in the EU?* (Question QA17b). As with the previous variable, each answer is assigned a numerical value and the (reverted) scale goes from 1 (not at all satisfied) to 5 (very satisfied).

EU knowledge is based on answers to the following question: *For each of the following statements about the EU, could you please tell me whether you think it is true or false?* (Question QA15). The statements about which the respondents should give their opinions relate to the following dimensions: 'The Euro Area currently consists of 19 Members'; 'The members of the European Parliament are directly elected by the citizens of each Member'; and 'Switzerland is a Member State of the EU'. Possible answers to these three statements are 'true', 'false' and 'don't know'. The scale that measures respondents' knowledge goes from zero (which equals the value of 1) to all three correct answers (value of 4).

Turning to the macro-level, the variable Energy Dependence gauges the extent to which an economy depends on imported energy and is calculated as the quotient of net imports of energy divided by the gross available energy (see Eurostat 2019). The greater the values, the greater the energy dependence and the greater a country's level of problem pressure. For the climate change and renewable energy outcome variables, we rely on the Climate Change Performance Index (CCPI) for the year 2018, which assesses the country's aggregated performance based on 14 indicators within the following four categories: greenhouse gas emissions, renewable energy, energy use, and climate policy (Burck et al. 2017). The greater the values, the better the country's performance and the lower the level of problem pressure. For the outcome variable that refers to environmental protection, we rely on the Environmental Performance Index (EPI) as measured in 2016. The EPI is based on 24 performance indicators across 10 issue categories, covering ecosystem vitality and environmental health (Hsu et al. 2016). The gauge is represented as a single value for each country and, as with the CCPI, higher values indicate better performances.

Lastly, we differentiate between West European and Central-East European member states. The variable labelled as CEE is coded 0 for the West European countries and 1 for Central-East European countries.

Table 1 gives an overview of the summary statistics of the variables that will enter the analysis in the next section. In this context, it should be noted that 21% of the respondents indicated that the priority of the Energy Union should be ensuring the security of energy supply. Only 15% of the respondents indicated that guaranteeing the EU's independence in the field of energy should be one of the priorities. As for environmental protection and the fight against climate change, the share of respondents goes up to 40% and 30% respectively. Lastly, around 41% of the respondents indicated that the top priority of the Energy Union should be the development of renewable energy. Therefore, we can see that more respondents would like to see the Energy Union prioritise climate and environmental protection than the issue of energy security.

Table 1. Summary statistics.

Variable	N	Mean	Std. Dev.	Min	Max
Outcome variables					
Renewable Energy	27,988	0.415	0.493	0	1
Environmental Protection	27,988	0.394	0.488	0	1
Climate Protection	27,988	0.305	0.460	0	1
Energy Security	27,988	0.210	0.407	0	1
Energy Independence	27,988	0.145	0.352	0	1
Individual-level variables					
Ideology	23,032	5.252	2.156	1	10
Problem Environment	27,988	0.723	0.260	0	1
Problem Energy Security	27,988	0.041	0.198	0	1
Problem Climate Change	27,988	0.116	0.320	0	1
EU Image	27,575	3.232	0.914	1	5
EU Democracy	25,030	2.556	0.742	1	4
Country Democracy	27,374	2.564	0.818	1	4
EU Knowledge	27,988	2.726	0.904	1	4
Gender	27,988	1.542	0.498	1	2
Age	27,988	3.101	0.995	1	4
Country-level variables					
EPI	27,988	85.994	2.769	80.150	90.680
CCPI	27,988	55.174	8.465	38.740	74.320
Energy Dependence	27,988	53.972	22.424	4.07	102.940
CEE	27,988	0.404	0.491	0	1

The method of choice for this analysis of our data is multi-level logit models with random intercepts. The reason for this choice is twofold. We analyse five outcome variables with a restricted value range (they can only take the values 0 and 1), which justifies the choice of logit models. On the other hand, the outcome, the focal explanatory and the control variables are measured both at the individual level and the country level, which explains why a multi-level approach is preferable to a single-level approach.

5. Empirical findings

We now turn to the findings of the multi-level logit models fitted to our outcome variables, which are presented in Table 2. The table comprises of five models, which only vary in respect of their outcome variables; fitting models that are identical increases the comparability of the findings and affords a hard test of the hypotheses. At the bottom of the table, we report the number of observations for individuals and the number of countries. It should be noted that the number of countries equals 30 because the data are reported separately for East and West Germany and for Great Britain and Northern Ireland.

We begin the discussion with the findings of the individual-level focal explanatory variables. When inspecting the table, we can discern that Ideology produces significant odds ratios across all five models, which confirms hypotheses H1a to H1c. For models 1 and 2 the odds ratios are greater than 1, but for models 3–5 they are smaller than 1. Consequently, a move of one unit towards the right end of the ideology scale increases the odds of indicating energy security or energy independence by about 4%. Conversely, a move of one unit towards the right reduces the odds of giving priority to climate protection, environmental protection, and renewable energy by about 2% and 3% respectively.

For a more straightforward interpretation of the effect sizes, we visualise in Figure 2 how the predicted probabilities change across the different values of ideology for indicating that the Energy Union should give priority to fighting climate change or increasing energy security. Regarding the climate-dimension, an individual with the most left-wing ideology has a predicted probability of about 35% of indicating this as the policy priority. An individual with the most right-wing ideology has only a predicted probability of about 29% of supporting that policy goal as a priority. For energy security, we can observe the reverse pattern in changes in the predicted probabilities. Most left-wing individuals have a predicted probability of about 18% of indicating this as the policy priority of the Energy Union, but the probability goes up to about 24% for the most right-wing respondents. If we compare the effect size of ideology, we can see that it is greater for the outcome variable on fighting climate change than for the one on energy security.

With regard to Problem Perception, individuals who regard environmental degradation as an important problem are more likely to prioritise environmental protection in the Energy Union, whereas individuals concerned with climate change have greater odds of expressing a preference for climate protection and the development of renewable energy. Likewise, individuals who consider energy security a problem have greater odds of naming energy independence as a priority of the Energy Union. Interestingly, the odds ratios of this covariate are statistically insignificant for energy security, though the direction of the effect corresponds to the reasoning underlying hypothesis H2a. However, it should also be noted that the odds ratios for energy independence are significant and

Table 2. Multilevel logit models.

	Model 1 Security	Model 2 Independence	Model 3 Climate	Model 4 Renewables	Model 5 Environment
Ideology	1.042	1.037	0.969	0.978	0.967
	(0.00835)***	(0.00940)***	(0.00716)***	(0.00671)**	(0.00664)***
Problem Environment	0.779	0.764	1.476	1.093	1.630
	(0.0535)***	(0.0600)***	(0.0802)***	(0.0585)	(0.0857)***
Problem Energy Security	1.111	1.323	0.847	1.212	0.871
	(0.0890)	(0.113)**	(0.0651)*	(0.0838)**	(0.0622)
Problem Climate Change	0.734	0.793	2.210	1.217	1.472
	(0.0413)***	(0.0499)***	(0.0970)***	(0.0532)***	(0.0635)***
EU Image	1.015	1.092	1.060	1.076	1.000
	(0.0202)	(0.0251)***	(0.0193)**	(0.0183)***	(0.0171)
Country Democracy	1.031	0.963	0.968	1.018	1.007
	(0.0246)	(0.0259)	(0.0207)	(0.0205)	(0.0203)
EU Democracy	0.944	1.061	1.025	1.041	1.063
	(0.0252)*	(0.0326)	(0.0251)	(0.0239)	(0.0246)**
EU Knowledge	1.090	1.287	1.039	1.123	0.962
	(0.0224)***	(0.0308)***	(0.0192)*	(0.0195)***	(0.0166)*
Age	1.151	0.969	0.923	0.869	0.937
	(0.0212)***	(0.0195)	(0.0147)***	(0.0130)***	(0.0140)***
Gender	0.880	0.732	1.274	0.943	1.232
	(0.0303)***	(0.0287)***	(0.0397)***	(0.0274)*	(0.0358)***
CEE	1.149	1.242	0.498	0.472	0.676
	(0.190)	(0.212)	(0.0668)***	(0.0669)***	(0.0916)**
Energy Dependence	0.995	0.998	0.995	0.995	1.000
	(0.00338)	(0.00352)	(0.00273)	(0.00288)	(0.00276)
EPI	0.987	0.970	1.016	1.039	1.018
	(0.0294)	(0.0297)	(0.0243)	(0.0262)	(0.0245)
CCPI	0.997	0.999	1.004	1.001	1.001
	(0.00956)	(0.00992)	(0.00780)	(0.00822)	(0.00785)
_cons	0.592	1.335	0.112	0.0434	0.150
	(1.464)	(3.402)	(0.224)	(0.0910)	(0.300)
Observations	20,992	20,992	20,992	20,992	20,992
Cases	30	30	30	30	30
AIC	21,406.1	17,619.9	24,876.7	27,508.0	27,492.6

Exponentiated coefficients; standard errors in parentheses.* $p < 0.05$, ** $p < 0.01$, *** $p < 0.001$

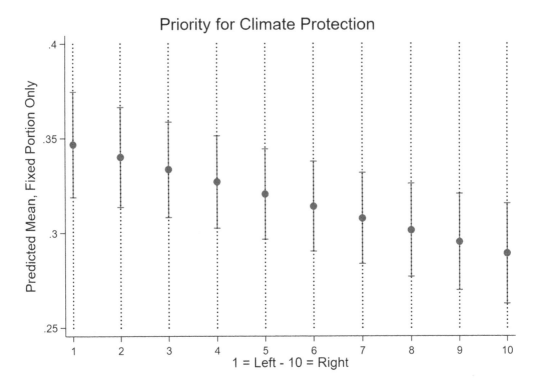

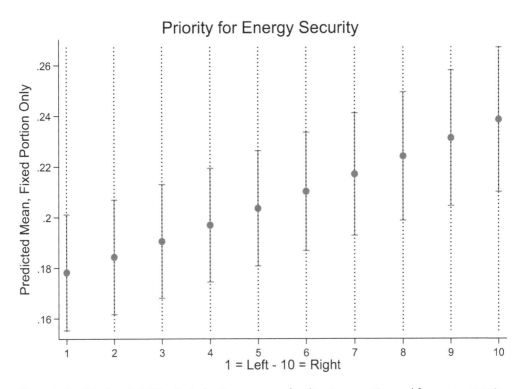

Figure 2. Predicted probabilities for indicating a priority for climate protection and for energy security.

smaller than 1, as postulated in H2a. Overall, we can see that the odds ratios of perceiving environmental degradation and climate change are greater than 1 for all outcome variables besides energy security and independence. In sum, we can confirm hypotheses H2a-2c.

Turning to the country-level focal explanatory variables, we can see that the odds ratios of respondents from Central-East European countries are significant and less than 1 in models 3–5. This means that Central-East Europeans are less likely to consider environmental protection or the fight against climate change necessary priorities of the Energy Union, a finding which supports hypotheses H3b and H3c. However, the odds ratios are not significant in models 1 and 2, which means that we have to reject H3a. Likewise, the variables gauging the problem-solving performance of the national governments fail to produce significant odds ratios in any of the models, and so hypotheses H4a-H4c must be rejected.

Turning to the control variables, we can see that in three models EU Image produces significant odds ratios greater than 1, indicating that respondents with a more positive perception of the EU are also more likely to indicate the development of renewable energy, the fight against climate change, and energy independence as policy priorities of the Energy Union. The variable EU Democracy produces odds ratios greater than 1 for all models besides the one on energy security. We can observe the same inconsistency in the direction of the effects for EU Knowledge, whereas the second outcome variable, environmental protection, is associated with odds ratios smaller in value than 1. The odds ratios also vary across the outcome variables for Age and Gender, but these findings are more plausible than those of the previous two control variables. Effectively, older people are less likely to indicate the environment- and climate-dimensions as the priority areas of the Energy Union than the security-dimension. Likewise, women have odds ratios greater than 1 of giving priority to the climate- and environment-dimension and odds ratios less than 1 for indicating the security-dimension as a priority field, although the finding for the development of renewable energy also deviates from this pattern.

6. Conclusion

European citizens are willing to confer more authority on the EU in the sense that they are in favour of a common energy policy. And they have specific preferences concerning the policy priorities of the Energy Union. The majority of respondents prefers the EU to develop renewable energy, arguably predominantly as a response to climate change. Environmental issues related to energy production and the fight against climate change also feature prominently with Europeans. Interestingly, the issue of energy security was named by a notably smaller group of respondents. As the empirical analysis has shown, a divide exists between West European and Central-East European citizens concerning their preferences for the policy priority of the Energy Union. The latter place less emphasis on the environment- and climate-dimensions of energy policy. In addition to this finding, we could show that the respondents' preferences were influenced by their ideological orientation and their perception of problems related to climate change, environmental degradation and energy security. In sum, we could confirm three of the four sets of hypotheses put forward and had to reject those addressing the problem-solving performances of national governments.

Connecting this study to the introductory article to this Special Issue (Herranz-Surrallés, Solorio, and Fairbrass forthcoming) as well as its theme more generally, we offered a complementary perspective on politicisation in the sense that we have revealed a European public that is more willing to delegate policymaking competences to the EU than to their respective national governments. The data consulted suggest that the governments of member states might be able to benefit electorally from the delegation of authority to the EU. In so doing, however, they need to be aware of the more specific mandate provided to them by their own citizens, which in most member states refers to the promotion of renewable energy. Nonetheless, it should be noted that the development of renewable energy can also help to strengthen energy security in the short run by diversifying the energy mix and in the long run by replacing fossil fuels (which often need to be imported) with renewables. Therefore, a viable strategy for a positive politicisation of renewable energy across all EU member states could be to stress more explicitly the multiple benefits of renewable energy, which also includes an increase in the security of energy supply.

Acknowledging the limits of the analyses presented in this study, we propose moving forward by using more refined measurements at the individual level. For example, it could prove fruitful to include information on whether the respondents personally experienced a shortage in energy supply and what the reason for this was. Our analysis did not sufficiently explore the salience of energy issues to the individual since no corresponding question was asked in the Eurobarometer. Therefore, it would be desirable to include this information in future research.

Overall, we believe that concentrating on the willingness of EU citizens to delegate energy policy competences to the EU is an insightful perspective, both for advancing the state of knowledge and for heightening our ability to inform policymaking. Research has alluded to the interrelationships between climate, energy and environmental policy, but it has paid little attention to their interconnection with regard to how citizens perceive them. If the EU wants to attain further integration in the domain of energy policy, we advise taking public opinion on the various dimensions of energy policy into account. The potential merit of the EU's policymaking competences needs to be highlighted to citizens if the EU wishes to increase its policymaking power, which it intends to do by means of the Energy Union. If the citizens understand the potential merit, they can put pressure on their respective governments to agree to future integration steps concerning energy policy. In this way, the EU may even obtain a more robust mandate for energy governance when reaching out to citizens.

Acknowledgments

We thank the guest editors, Isabelle Stadelmann-Steffen and an anonymous reviewer for constructive comments on earlier drafts. Laurence Crumbie deserves credit for language editing.

Disclosure statement

No potential conflict of interest was reported by the authors.

Funding

This work was supported by the Heidelberg Center for the Environment and the Konrad Adenauer Foundation.

ORCID

Mile Mišić ⓘ http://orcid.org/0000-0002-6554-1962

References

Alexandrova, P., M. Carammia, S. Princen, and A. Timmermans. 2014. "Measuring the European Council Agenda: Introducing a New Approach and Dataset." *European Union Politics* 15 (1): 152–167. doi:10.1177/1465116513509124.

Álvarez-Verdugo, M. 2015. "The EU 'stress Tests': The Basis for a New Regulatory Framework for Nuclear Safety." *European Law Journal* 21 (2): 161–179. doi:10.1111/eulj.12114.

Austvik, O. G. 2016. "The Energy Union and Security-of-gas Supply." *Energy Policy* 96: 372–382. doi:10.1016/j.enpol.2016.06.013.

Biedenkopf, K., and F. Petri. 2019. "EU Delegations in European Union Climate Diplomacy: The Role of Links to Brussels, Individuals and Country Contexts." *Journal of European Integration* 41 (1): 47–63. doi:10.1080/07036337.2018.1551389.

Boasson, E. L., and J. Wettestad. 2013. *EU Climate Policy: Industry, Policy Interaction and External Environment*. Abindgdon: Routledge.

Bocquillon, P., and M. Dobbels. 2014. "An Elephant on the 13th Floor of the Berlaymont? European Council and Commission Relations in Legislative Agenda Setting." *Journal of European Public Policy* 21 (1): 20–38. doi:10.1080/13501763.2013.834548.

Bocquillon, P., and T. Maltby. forthcoming. "EU Energy Policy Integration as Embedded Intergovernmentalism: The Case of Energy Union Governance Regulation". *Journal of European Integration* this issue.

Bouzarovski, S., and S. Tirado Herrero. 2017. "The Energy Divide: Integrating Energy Transitions, Regional Inequalities and Poverty Trends in the European Union." *European Urban and Regional Studies* 24 (1): 69–86. doi:10.1177/0969776415596449.

Braun, D., and M. Tausendpfund. 2014. "The Impact of the Euro Crisis on Citizens' Support for the European Union." *Journal of European Integration* 36 (3): 231–245. doi:10.1080/07036337.2014.885751.

Budge, I. 2013. "The Standard Right-left Scale." Colchester: Essex University. accessed 12 November 2019. https://manifestoproject. wzb. eu/down/papers/budge_right-left-scale.pdf

Burck, J., F. Marten, C. Bals, N. Höhne, C. Frisch, N. Clement, and K. Szu-Chi. 2017. "Climate Change Performance Index. Results 2018." Bonn: Germanwatch. accessed 12 November 2019 https://www.climate-change-performance-index.org/sites/default/files/documents/the_climate_change_performance_index_2018.pdf

Cherp, A., and J. Jewell. 2014. "The Concept of Energy Security: Beyond the Four As." *Energy Policy* 75: 415–421. doi:10.1016/j.enpol.2014.09.005.

Clements, B. 2012. "Exploring Public Opinion on the Issue of Climate Change in Britain." *British Politics* 7 (2): 183–202. doi:10.1057/bp.2012.1.

Eberlein, B. 2012. "Inching Towards a Common Energy Policy: Entrepreneurship, Incrementalism, and Windows of Opportunity." Chap. 8 In *Constructing a Policy-Making State? Policy Dynamics in the EU*, edited by J. Richardson, 147–169, Oxford: Oxford Scholarship Online. doi:10.1093/acprof:oso/9780199604104.003.0008

European Commission. 2015. Communication from the Commission to the European Parliament, the Council, the European Economic and Social Committee, the Committee of the Regions and the European Investment Bank: A Framework Strategy for A Resilient Energy Union with

A Forward-Looking Climate Change Policy/* COM/2015/080 Final */." Brussels: European Commission. https://eur-lex.europa.eu/resource.html?uri=cellar:1bd46c90-bdd4-11e4-bbe1-01aa75ed71a1.0001.03/DOC_1&format=PDF

European Commission. 2018a. Standard Eurobarometer 89 Spring 2018: Report. Brussels: European Commission. doi:10.2775/00.

European Commission: Eurobarometer 89.1. 2018b. "TNS Opinion & Social, Brussels Producer." (ZA 6963 data file version 1.0.0). Cologne: GESIS Data Archive. doi:10.4232/1.13154.

European Council. 2006. *Conclusions of the European Council*. 23–24 March. http://data.consilium.europa.eu/doc/document/ST-7775-2006-INIT/en/pdf

Eurostat. 2019. "Energy Dependence Table." Eurostat. https://ec.europa.eu/eurostat/databrowser/view/t2020_rd320/default/table?lang=en

Hadjar, A., and M. Beck. 2010. "Who Does Not Participate in Elections in Europe and Why Is This? A Multilevel Analysis of Social Mechanisms behind Non-voting." *European Societies* 12 (4): 521–542. doi:10.1080/14616696.2010.483007.

Herranz-Surrallés, A. 2016. "An Emerging EU Energy Diplomacy? Discursive Shifts, Enduring Practices." *Journal of European Public Policy* 23 (9): 1386–1405. doi:10.1080/13501763.2015.1083044.

Herranz-Surrallés, A. 2017. "Energy Diplomacy under Scrutiny: Parliamentary Control of Intergovernmental Agreements with Third-country Suppliers." *West European Politics* 40 (1): 183–201. doi:10.1080/01402382.2016.1240406.

Herranz-Surrallés, A., I. Solorio, and J. Fairbrass. forthcoming. "Renegotiating Authority in the Energy Union: A Framework for Analysis". *Journal of European Integration* this issue.

Hsu, A., D. Esty, M. Levy, A. de Scherbinin. 2016. "The 2016 Environmental Performance Index Report." New Haven, CT: Yale Centre for Environmental Law and Policy. https://issuu.com/2016yaleepi/docs/epi2016_final

Kumlin, S., and I. Stadelmann-Steffen. 2014. "Citizens, Policy Feedback, and European Welfare States." In *How Welfare States Shape the Democratic Public. Policy Feedback, Participation, Voting and Attitudes*, edited by S. Kumlin and I. Stadelmann-Steffen, 3–16. Cheltenham: Edward Elgar. doi:10.4337/9781782545491.00007

Kustova, I. 2017. "Towards a Comprehensive Research Agenda on EU Energy Integration: Policy Making, Energy Security, and EU Energy Actorness." *Journal of European Integration* 39 (1): 95–101. doi:10.1080/07036337.2017.1258757.

Latré, E., P. Thijssen, and T. Perko. 2019. "The Party Politics of Nuclear Energy: Party Cues and Public Opinion regarding Nuclear Energy in Belgium." *Energy Research & Social Science* 47: 192–201. doi:10.1016/j.erss.2018.09.003.

Maggetti, M., and P. Trein. 2019. "Multilevel Governance and Problem-solving: Towards a Dynamic Theory of Multilevel Policy-making?" *Public Administration* 97 (2): 355–369. doi:10.1111/padm.12573.

Maltby, T. 2013. "European Union Energy Policy Integration: A Case of European Commission Policy Entrepreneurship and Increasing Supranationalism." *Energy Policy* 55: 435–444. doi:10.1016/j.enpol.2012.12.031.

McCright, A. M., R. E. Dunlap, and S. Marquart-Pyatt. 2016. "Political Ideology and Views about Climate Change in the European Union." *Environmental Politics* 25 (2): 338–358. doi:10.1080/09644016.2015.1090371.

Mettler, S., and J. Soss. 2004. "The Consequences of Public Policy for Democratic Citizenship: Bridging Policy Studies and Mass Politics." *Perspectives on Politics* 2 (1): 55–73. doi:10.1017/S1537592704000623.

Mišík, M. 2015. "The Influence of Perception on the Preferences of the New Member States of the European Union: The Case of Energy Policy." *Comparative European Politics* 13 (2): 198–221. doi:10.1057/cep.2013.9.

Neumayer, E. 2004. "The Environment, Left-wing Political Orientation and Ecological Economics." *Ecological Economics* 51 (3–4): 167–175. doi: 10.1016/j.ecolecon.2004.06.006

Pointvogl, A. 2009. "Perceptions, Realities, Concession. What Is Driving the Integration of European Energy Policies?" *Energy Policy* 37 (12): 5704–5716. doi:10.1016/j.enpol.2009.08.035.

Ringel, M., and M. Knodt. 2018. "The Governance of the European Energy Union: Efficiency, Effectiveness and Acceptance of the Winter Package 2016." *Energy Policy* 112: 209–220. doi:10.1016/j.enpol.2017.09.047.

Solorio, I. 2011. "Bridging the Gap between Environmental Policy Integration and the EU's Energy Policy: Mapping Out the 'green Europeanisation' of Energy Governance." *Journal of Contemporary European Research* 7 (3): 396–415.

Solorio, I., and F. Morata. 2012. "Introduction: The Re-evolution of Energy Policy in Europe." Chap. 1 In *European Energy Policy: An Environmental Approach*, edited by I. Solorio and F. Morata, 1–22. Cheltenham: Edward Elgar. doi: 10.4337/9780857939210.00008

Solorio, I., and H. Jörgens. forthcoming. "Contested Energy Transition? Europeanization and Authority Turns in EU Renewable Energy Policy". *Journal of European Integration* this issue.

Stadelmann-Steffen, I., and C. Eder. 2020. " Public opinion in policy contexts. A comparative analysis of domestic energy policies and individual policy preferences in Europe". *International Political Science Review*, forthcoming.

Szulecki, K. 2016. "European Energy Governance and Decarbonization Policy: Learning from the 2020 Strategy." *Climate Policy* 16 (5): 543–547. doi:10.1080/14693062.2016.1181599.

Szulecki, K., and K. Westphal. 2018. "Taking Security Seriously in EU Energy Governance: Crimean Shock and the Energy Union." Chap. 7 In *Energy Security in Europe*, edited by K. Szulecki, 177–202. Cham: Palgrave Macmillan. doi:10.1007/978-3-319-64964-1_7

Szulecki, K., S. Fischer, A. T. Gullberg, and O. Sartor. 2016. "Shaping the 'energy Union': Between National Positions and Governance Innovation in EU Energy and Climate Policy." *Climate Policy* 16 (5): 548–567. doi:10.1080/14693062.2015.1135100.

Toke, D., and S. E. Vezirgiannidou. 2013. "The Relationship between Climate Change and Energy Security: Key Issues and Conclusions." *Environmental Politics* 22 (4): 537–552. doi:10.1080/09644016.2013.806631.

Tosun, J., and I. Solorio. 2011. "Exploring the Energy-environment Relationship in the EU: Perspectives and Challenges for Theorizing and Empirical Analysis." *European Integration Online Papers* 15 (1): 1–15. doi:10.1695/2011007

Tosun, J., L. Zöckler, and B. Rilling. 2019. "What Drives the Participation of Renewable Energy Cooperatives in European Energy Governance?" *Politics and Governance* 7 (1): 45–59. doi:10.17645/pag.v7i1.1782.

Wettestad, J., P. O. Eikeland, and M. Nilsson. 2012. "EU Climate and Energy Policy: A Hesitant Supranational Turn?" *Global Environmental Politics* 12 (2): 67–86. doi:10.1162/GLEP_a_00109.

Appendix

Table A1. Items from the Eurobarometer survey used in this study.

Item	Item in the questionnaire	Short description	Question	Question mode
Outcomes	QB2	*Policy preference*	In your opinion, which of the following objectives should be given top priority in a European energy union? (a) Guaranteeing reasonable energy prices for consumers (b) Guaranteeing a continuous supply of energy (c) Guaranteeing EU's independence in the field of energy (d) Protecting the environment (e) Interconnecting energy infrastructure (f) Fighting global warming (g) Guaranteeing the competitiveness of EU's industry (h) Reducing energy consumption (i) Guaranteeing reasonable prices for companies (j) Pooling EU's negotiating power towards energy providers (k) Developing renewable energy (l) Other (m) None (n) Don't know	Selection of maximum three answers
1	QA15	*Knowledge about the EU*	For each of the following statements about the EU, could you please tell me whether you think it is true or false? (a) The Euro area currently consists of 19 Members (b) The members of the European Parliament are directly elected by the citizens of each Member (c) Switzerland is a Member State of the EU	(1) True, (2) False, or (3) Don't know
2	QA9	*Perception of the EU*	In general, does the EU conjure up for you a positive, fairly positive, neutral, fairly negative or very negative image?	Scale from Very positive (1) to Very negative (5), including Don't know answer.
3	QA17a	*Satisfaction with democracy in the country*	On the whole, are you very satisfied, fairly satisfied, not very satisfied or not all satisfied with the way democracy works in our country?	Scale from Very satisfied (1) to Not at all satisfied (4), including Don't know answer.
4	QA17b	*Satisfaction with democracy in the EU*	And how about the way democracy works in the EU?	Scale from Very satisfied (1) to Not at all satisfied (4), including Don't know answer.
5	D1	*Ideological self-placement*	In political matter people talk of "the left" and "the right". How would you place your views on this scale?	Scale from Left (1) to Right (10). Includes Refusal and Don't know as options.
6	QA5	*Issue perception*	What do you think are the two most important issues facing the EU at the moment? (a) Crime (b) Economic situation	Selection of maximum three answers.

(*Continued*)

Table A1. (Continued).

Item	Item in the questionnaire	Short description	Question	Question mode
			(c) Rising prices/inflation/cost of living	
			(d) Taxation	
			(e) Unemployment	
			(f) Terrorism	
			(g) EU's influence in the world	
			(h) The state of Member States' public finances	
			(i) Immigration	
			(j) Pensions	
			(k) The environment	
			(l) Energy supply	ffigur
			(m) Climate change	
			(n) Other	
			(o) None	
			(p) Don't know	

EU energy policy integration as embedded intergovernmentalism: the case of Energy Union governance

Pierre Bocquillon and Tomas Maltby (iD)

ABSTRACT
The launch of the 'Energy Union' in 2014, represented a major step to deepen EU cooperation in energy and climate policies. Yet, in energy, member states have remained particularly jealous of their sovereignty, limiting the pace and scope of integration. EU energy policy appears to fit the specifications of 'new intergovernmentalism' (NI). Member states have been keen on reinforcing cooperation but have refrained from delegating further authority to supranational institutions, preferring to maintain a high level of control within the Council and European Council. However, focusing on the Energy Union Governance Regulation adopted in 2018, we argue that the sector does not fit neatly within this NI account. Although governments remain central to the process, they operate within a hybrid institutional framework combining supranational and intergovernmental elements, in which formal and informal authority distribution is unstable and contested. We suggest this form of governance is better described as 'embedded intergovernmentalism'.

Introduction

In the past decade and a half, energy policy has moved from a marginal position to a prominent and dynamic area of European integration. Energy issues have been increasingly harnessed to climate policy and have become particularly salient internationally within the context of international climate negotiations and the urgency of an energy transition to meet Paris Agreement commitments. In Europe energy policy is also perceived as an area with the potential to reinvigorate the integration process and demonstrate the 'added value' of the EU to its member states and citizens (e.g. Andoura et al. 2010). The 'Energy Union', launched by the European Commission in 2014, is a high-profile initiative to deepen cooperation in this area.

Since its inception, EU energy policy has developed without a clearly delineated legal foundation. It has emerged out of the EU's competencies in contiguous policy areas – particularly the internal market and environmental provisions (Morata and Solorio 2012). This has meant an important role for the European Commission and the Court of Justice of the EU (CJEU). On the other hand, member states have remained particularly guarded when it comes to their sovereignty in this area, limiting the pace and scope of integration (Slominski 2016, 345). Since the entry into force of the Lisbon Treaty, the EU has formal competencies in energy. Community competencies now explicitly encompass the internal energy market, as well as clean energy and the security of energy supplies. Yet legal authority remains firmly with member states in significant areas such as national resources, the ability to determine national energy mixes, as well as taxation. Member state governments have also been at the centre of the decision-making process, within the Council and increasingly through the European Council which has significant authority to decide key energy issues such as climate and energy targets. On this account, EU energy policy appears to fit the specifications of the so-called 'new intergovernmentalism' post-Maastricht, which sees member states as keen on reinforcing cooperation in sovereignty sensitive areas while refraining from delegating further powers to supranational authorities.

This paper examines the EU's Energy Union governance framework to answer two related questions: 1. Does the evolution of EU energy governance as part of the 'Energy Union' project continue or depart from pre-existing patterns of authority distribution? 2. Does the new EU energy and climate governance framework fit the new intergovernmentalist model of authority distribution? We examine how the key actors involved approach the distribution of authority between member states and EU institutions, and what the negotiations and their outcome tell us about where authority lies in energy policy, how this distribution is contested, and how contestation is managed in the sector.

We argue that different legal bases and roles for supranational institutions mean that the boundaries of energy policy are unclear, increasingly overlapping with climate policy, and the distribution of authority is unstable and contested. It is a hybrid area combining intergovernmental and Community modes of governance. Although member states exercise significant authority through the Council and increasingly the European Council, this takes place within the context of the EU's institutional environment.

The first section reviews critically the new intergovernmentalist agenda, discusses the limits of its application to EU energy policy, and contrasts three governance models – new intergovernmentalism, the Community method and a third hybrid model we call 'embedded intergovernmentalism' – focusing on three dimensions – decision-making and decisions; the role of supranational institutions (notably in implementation); and compliance mechanisms. This framework is then applied to Energy Union governance, contrasting it with the pre-existing governance framework to identify and characterise shifting patterns of authority. We focus in particular on the 2018 Energy Union Governance Regulation, a central piece of the 2030 clean energy governance framework. The following section reflects on patterns of authority distribution in the Energy Union and best ways to conceptualise energy policy governance. We conclude that a better way to understand EU energy policy integration is to conceptualise it as a form of 'embedded intergovernmentalism', in which governments remain central and particularly jealous of national prerogatives, while being deeply entrenched in the EU institutional framework.

The community method, new intergovernmentalism or embedded intergovernmentalism?

The 'new intergovernmentalism' debate

The new intergovernmentalism (NI) that has emerged after the Maastricht Treaty is characterised by 'integration without supranationalism' (Bickerton, Hodson, and Puetter 2015a, 703, 2015b). Member states have shown an appetite for enhanced cooperation while being simultaneously reluctant to delegate further sovereignty and bring sensitive areas within the realm of the traditional Community method of EU decision-making. This has been most evident in the new areas of integration that have developed after Maastricht, namely the common foreign and security policy, economic and monetary union, as well as parts of justice and home affairs and social and employment policies. Here governments have privileged the search for consensus through intergovernmental fora as well as the delegation of powers to 'de novo bodies', rather than to supranational institutions (Bickerton et al. 2015a). It is not clear, however, whether NI is meant to characterise only the so-called 'new' areas of integration or the European project as a whole, including the former first pillar of the Maastricht Treaty. As Bulmer argues (Bulmer 2015, 294), NI rightly points to an intergovernmental tendency in European integration, but integration post-Maastricht also includes moves towards supranationali-sation in certain areas.

Integration without supranationalism is attributed to changes in the political economy of member states since the 1980s with the end of the post-war Keynesian consensus, as well as change in state-society relations and increasing tensions between elites and citizens over integration (Bickerton 2012). Indeed, the growth of domestic contestation has created a form 'constraining dissensus' that limits the possibilities of functional integration (Hooghe and Marks 2009), to which intergovernmental cooperation is a response, redirecting rather than redressing legitimacy problems (Hodson and Puetter 2019).

NI is also associated with a micro theory of institutional change that Puetter (2012, 2014) has labelled 'deliberative intergovernmentalism'. In sensitive areas, national government are eager to seek commonly agreed solutions while remaining anxious of being by-passed and adamant about preserving their veto rights. This translates into the preference for consensus, which intergovernmental institutions such as the Council, European Council and Eurogroup are geared towards producing. This consensus is engineered through specific institutional features (such as more regular meetings, informality, secrecy) which aim to favour 'honest' deliberation, problem solving and cooperation (Maricut and Puetter 2018). NI emphasises and explains consensus-seeking between European governments but, as a result, tends to overlook enduring divisions and conflict (Schimmelfennig 2015, 728) and 'largely disregard[s] the significance of coercive or institutional power' (Schmidt 2018, 1546). As Novak (2013) contends, consensus can also be 'apparent' rather than actual, as deliberation can hide latent conflict and power relations.

Energy as a new intergovernmental area?

Energy is an area in which national traditions and prerogatives are deeply rooted. Member states' approaches and preferences to a large extent reflect their national resources, energy mixes and the structures of their national energy systems, as well as their political histories. Energy policies are still perceived as key to national sovereignty and even national security (Judge and Maltby 2017), and national governments are reluctant to cede authority to the supranational level on issues that have direct and important consequences for their citizens' welfare.

As a result, the upward delegation of authority to the EU level and supranational institutions has been limited and belated. At the beginning of the 1990s the energy sector was one of the least integrated sectors in the European Communities (Matlary 1996). However, since the mid-1990s, despite the lack of firm legal basis, energy policy cooperation has deepened within the framework of the traditional Community method, with the Commission as (formally) the sole initiator of legislation, the European Parliament (EP) as co-legislator and a role for the Court in adjudicating disputes and enforcement (see Dehousse 2011). The energy sector was included in the internal market agenda in the late 1990s (Eising 2002) and affected by the growing body of environmental legislation via the EU's wide-ranging competences in that area (Morata and Solorio 2012). The Commission has been instrumental in using its legal prerogatives in other sectors, in particular the internal market (including competition rules) to push integration forward in energy (Schmidt 1998). These developments were eventually codified in the Lisbon Treaty (art. 194) and expanded to security of supply measures, the Treaty explicitly specifying that national resources, the energy mix and taxation remain strictly national prerogatives – a reflection of member states' enduring sovereignty concerns. However, the importance of article 194 should not be overstated as it has simply institutionalised what had developed progressively in practice (Piris 2010, 319).

Decision-making and decisions

Energy negotiations have been highly contested and member states have been at the forefront both individually and collectively (e.g. Buchan 2009). The European Council has gained influence due to the need for inter-sector coordination at the highest level to set mid- to long-term energy and climate objectives (such as the 2020 and 2030 targets) and national political authority to respond to energy crises. Thaler (2016) stresses the role of the European Council in energy policy development, as central to facilitating integration through ironing out dissensus. On this account, energy policy shares certain characteristics with the 'new areas' of inter-governmental cooperation identified by new intergovernmentalism. Member states have been keen on reinforcing cooperation in the area, while at the same time refraining from delegating authority to supranational institutions, preferring to maintain national sovereignty over key aspects. Intergovernmental fora, where consensus nominally prevails, have been central to this process.

However, Thaler's account overstates the European Council's capacity to steer the policy-making process and monitor implementation. The European Council meets relatively infrequently (usually quarterly) and discusses energy and climate policy only occasionally within these meetings. Its attention to these issues tends to peak after energy crises (e.g. the 2009 gas supply disruption) or in preparation for international climate conferences (e.g. 2007-8, 2013-4) (Alexandrova 2015). A new intergovernmentalist perspective also underplays that intergovernmental bargaining takes place within the parameters of a supranational framework (Stone Sweet and Sandholtz 1997, 299–300), which shapes its process and outcomes. In hybrid areas such as energy, we argue that there is a more systematic interplay between intergovernmental and legislative processes, with the former remaining central but embedded in the latter.

Role of supranational institutions

NI claims that post-Maastricht the Commission refrains from pursuing further integration as it is sensitive to member states' concerns (e.g. Hodson 2013), underlining the importance of ownership by member states through consensual decision-making at the top level to ensure the authority of the decisions and ensure smooth implementation and compliance (Puetter 2012). Yet, this perspective misses the influence of supranational institutions based on their legislative prerogatives and ability to influence intergovernmental debates. In hybrid governance areas such as energy, European Council conclusions do set the framework within which the legislators must work, but the European Commission plays a substantial role in framing policy debates and following up with legislative proposals (Bocquillon and Dobbels 2014), and implementation including the possibility of infringement proceedings if member states fail to comply. Its formal and informal agenda setting powers – exclusive right of legislative initiative, decision on the timing of proposals – provides the Commission with the ability to steer the process and shape outcomes. The European Parliament has been progressively empowered as a co-legislator in most areas of energy policy, acquiring institutional power in the decision-making process both in agenda setting and in co-decision by putting pressure on the Council but also through ideational influence and discursive interactions – in link with a 'new parliamentarist' view of EU integration (Schmidt 2018).

Compliance mechanisms

In areas dominated by NI, key decisions tend to be first and foremost of a political nature – often in the form of European Council conclusions and intergovernmental agreements, even though legislation is not absent. In comparison, in areas subject to the traditional Community method, implementation and compliance are ensured through legislative and judicial means. In sectors characterised by hybrid governance objectives are often non- (or partially) binding and the obligation falling on member states general in scope. Compliance mechanisms here tend to incorporate a process of reporting by governments, monitoring of progress by the Commission, combined with peer-pressure for those governments who are lagging behind. Given the EU's circumscribed but significant legal competence, there is a 'shadow of hierarchy' (Eberlein 2008). This takes the form of court sanctions or the potential of issuing 'harder', more constraining, legislation to help steer the implementation process and ensure compliance. Intergovernmental and supranational modes of governance are therefore intertwined, and interact with one another.

Case and methods

We look at the validity of NI claims and propose an 'embedded intergovernmentalist' approach for the case of the Energy Union governance framework. In this section we briefly present the contextual background to the case and methodology used, before presenting the empirical material and analysis.

Table 1. Patterns of authority distribution in three models of governance.

Modes of governance	New intergovernmentalism	Embedded intergovernmentalism (hybrid)	Community method
Decision-making & decisions	Consensual deliberation	Combination of legislative and non-legislative processes Political decisions set the framework and headline goals, within which legislation and non-legislative acts are inserted	Legislative decision-making and legislation
Role of supranational institutions	Commission refrains from pursuing further integration	Supranational institutions can influence and shape national positions but ultimately rely on MS decisions	Integrationist role for the Commission, EP and CJEU
Compliance mechanisms	Ownership of the decisions by national governments, through new and intergovernmental decisions	Reporting, monitoring and peer pressure in the shadow of legislation/the Court	Legal enforcement by the Commission and Court
Overall institutional power balance	Dominance of Council and EUCO	Council and EUCO central but inserted in institutional framework	Distributed through the 'institutional triangle'

Source: Based on Dehousse 2011; Bickerton et al. 2015b; and authors' elaboration.

Case study: the governance framework of the Energy Union

In April 2014 the Polish Prime Minister proposed an 'Energy Union', focused on supply security objectives, including a joint gas purchase mechanism to strengthen the hand of EU member states vis à vis external suppliers (Szulecki et al. 2016). The new Commission President Juncker seized on the Energy Union project as a priority in Autumn 2014 and created the post of Energy Union Vice President to drive it forward.

In preparation for the Paris Climate summit of December 2015, the Autumn 2014 European Council conclusions set out the 2030 climate and energy policy framework, including its three main targets: 27% energy efficiency (see Dupont 2020), 27% renewables and a 40% GHG emission reduction. It was agreed that: ' ... a reliable and transparent governance system without any unnecessary administrative burden will be developed to help ensure that the EU meets its energy policy goals' (European Council 2014, 9). As the proposed 2030 energy efficiency goal was non-binding and the renewable energy target was binding only at EU level, a governance mechanism was promoted – in particular by Germany, along with France and the Scandinavian countries – to ensure a collective effort by member states (Vandendriessche, Saz-Carranza, and Glachant 2017, 18).

The Commission's 2015 Energy Union strategy included five broad priority areas.[1] The initial breadth, or vagueness, of the concept made it hard to oppose for member states and industry stakeholders, particularly when there was support from the public for energy integration (interview 3), and it was endorsed by the European Council the following month (European Council 2015). The Commission's strategy emphasised the requirement for 'an integrated governance and monitoring process' (European Commission 2015, 17). A Regulation on Governance of the Energy Union was formally proposed as part of the November 2016 'Clean Energy package', composed of eight pieces of legislation, the other seven of which were revisions to existing directives and regulations.

The 2015 Paris climate negotiations saw the EU join the High Ambition Coalition, which pushed for limiting emissions to levels compatible with a 1.5°C global temperature increase (going further than the established 2°C target). Adopted in December 2015, the Paris Agreement includes the aspirational 1.5°C target and proposes a bottom-up process through which the parties define their own plans towards achieving the overall objective through Nationally Determined Contributions (NDCs), which should be reviewed and upgraded at regular intervals. The EU's commitment to the Paris Agreement has shaped its own governance mechanism (Oberthür 2019).

The Governance Regulation 'aims to integrate, simplify and align an overlapping set of planning, reporting and monitoring requirements ("obligations") under the existing EU energy and climate acquis' (Wilson 2018, 2; also Ringel and Knodt 2018), after the Commission concluded that there was a lack of policy coherence, efficiency and consistency between climate and energy policy fields (European Commission 2016). The aim is also to provide a robust process to monitor implementation and ensure that member states are on track to achieve EU headline goals and Paris Agreement commitments.

Methodology

The research takes a qualitative approach, using rich primary data to analyse authority conflicts and compromises regarding the Energy Union governance architecture. We focus in particular on the EU's 2018 Regulation on Governance of the Energy Union, and interconnected legislation proposed as part of the winter 2016 'Clean Energy Package'. The governance structure is discussed in comparison to pre-existing governance frameworks, in particular those established as part of the European Climate Change Programme in the early 2000s, and the 2009 climate and energy package. The qualitative approach includes the use of documentary evidence, press releases and eighteen elite interviews with representatives of the member states, the European Commission and Parliament, the Council Secretariat, energy regulators and industry representatives, and energy experts working in the media, think tanks and NGOs. They were selected based on their centrality in the EU-based energy policy network and the Clean Energy Package's policy-making process.

Negotiating the Energy Union governance framework

The empirical analysis is divided into three sections, as identified in our discussion of NI as applied to energy, and the theoretical framework: decision-making, the role of supranational institutions, and compliance mechanisms.

Decision-making: setting the EU's climate ambition

A key debate within the Clean Energy Package centred around the level and legally binding nature of the EU's climate ambition in the context of the Paris Agreement. In October 2014 the European Council, based on a Commission proposal, increased the 2020 targets of 20-20-20 (emissions reduction, renewables, energy efficiency improvements) to 40-27-27 by 2030. While as part of the 2020 Framework both the GHG and renewable targets were binding at the national level, in the 2030 Framework it is the case only for the GHG target.[2] Influenced by Energy Commissioner Oettinger, the Commission refrained from pushing for binding national commitments for renewable energy and energy efficiency (Bürgin 2015). It was criticised for failing to take an ambitious stance against reluctant member states, with the UK supported by Spain, Poland and other Central and Eastern European countries rejecting binding renewable targets as an infringement of their authority to decide their own energy mix (Nelsen 2014; interview 7). The level of ambition was generally considered disappointing for renewable energy and energy efficiency (van Renssen 2014). As assumed by NI, the Commission avoided more integrationist measures, while the European Council played a central role in defining the framework which the Commission had to work within when devising its clean energy package.

In 2017, the debate about climate ambition re-emerged, in a context where the EU wanted to demonstrate its ambition as part of the implementation of the 2015 Paris Agreement, while climate issues had also risen in the agenda of several member states. It exposed a power struggle between member states themselves, as well as with supranational institutions. During the negotiations in the Council, Germany and France worked closely together (interview 16), while Luxemburg, Portugal, The Netherlands and Sweden were part of the most ambitious coalition (interviews 8, 13, 17). The negotiation dynamics changed over time. Following elections in 2017, the German government ceased coordinating with ambitious states (interview 8).[3] Instead, France was now demonstrating climate leadership under its newly elected President and high-profile environment Minister Nicolas Hulot, along with Sweden and Luxembourg (interview 6). The increase in EU ambition, embodied in the targets, was significantly aided by changes in governments in Spain and Italy as negotiations neared an end. With Spain, this new ambition on climate and energy was espoused by the new socialist energy minister – leading a newly created Ministry for Ecological Transition (interviews 8; 11; 13; 15). The change in Italy's position, a result of the new Lega-Five Star Movement coalition government and driven mainly by the Five Star Movement's environmental commitments, was surprising, 'swinging the balance in favour of higher ambition' (interviews 8, 17). It led to the unravelling of a potential blocking minority at the June energy council, in the late stages of the trilogue with the European Parliament (interview 15). Whilst the Visegrad Group were opposed to ambitious renewable targets (interview 8) and a shift of authority to the EU level in this area (interview 12), Poland's likely opposition was constrained by hosting the UN's 2018 climate change conference (interview 15), and 'Bulgaria was neutralised by being the President' during the negotiations (interview 15). On Energy Efficiency there was similarly no blocking minority as the target was strictly non-binding. During the negotiations, authority delegation was therefore contested. Member states were not united in their preferences for the role of supranational institutions as assumed by NI, their positions shaped by changing national and international contexts.

It was within this context that the EP was able to influence the level of ambition. The Parliament was decisive in shifting ambitions upwards (interviews 8, 12, 16, 17, 18), seeking a 35% renewable target by 2030 instead of the original 27% (EP 2018), and a 40% instead of 27% target for energy efficiency. At the end of the negotiations, the Council and EP agreed on a compromise of increasing ambition to 32% for renewables and 32.5% for energy efficiency. On the nature of the targets, the EP was less successful, accepting that the renewable target would remain binding at EU level only and that the energy efficiency target was kept non-binding as preferred by member states. Therefore, the Council, while shifting its position on the levels of ambition, successfully resisted an upward shift of authority towards the EU level.

Another related and contested issue was the trajectory to be followed to reach the renewable energy targets. While the Commission had proposed a first check of national and EU progress towards these EU goals in 2023, member states agreed with the EP's proposal to have an earlier first check on progress, in 2022, but successfully countered that there should only be three (not four) reference points – to avoid administrative burden and inefficiency. Member states also pushed for an exponential rather than linear trajectory – as proposed by the Commission – towards nationally 'planned contributions'. Such an 'improvement focused approach' would require greater efforts close to 2030 to reduce costs by waiting for technology improvements (interview 1). Eventually a compromise agreement was adopted, which saw the Council and EP meet in the middle.[4]

In the context of the Paris Agreement, the EP has also pushed for the EU to commit to net zero emissions by 2050 (EP 2018).[5] Member states were divided, with France, Sweden and Luxembourg pushing for the net zero target but others opposed (Simon 2018a). The

Council agreed only that plans should be consistent with the Paris Agreement (CoEU 2018), however the European Council in March 2018 requested the Commission to prepare a long-term Paris-compliant climate strategy to be drafted in 2019 and finalised in 2020 (European Council 2018).[6] This aligned EP and ambitious member state preferences, and the delegated authority permitted the Commission to then strongly influence member states' climate ambition by framing it in terms of their Paris Agreement commitments.

The role of supranational institutions: planning and reporting

In the 2020 governance framework, EU and national targets were defined as part of different pieces of legislation (for more details on the 2020 renewable energy and energy efficiency targets see respectively: Solorio and Jörgens 2020; Dupont 2020). Member states had to establish separate plans for renewables, energy efficiency or non-ETS sector emissions, presenting the measures to be adopted to reach their sectoral objectives. In the new governance framework planning and reporting obligations are consolidated into integrated 'National Energy and Climate Plans' (NECPs). Member states define their national renewable and energy efficiency contributions towards EU targets as part of NECPs, in a bottom-up fashion based on the model of the Paris Climate agreement. This approach partially confirms NI's hypotheses about the dominance of national sovereignty concerns and preference for enhanced intergovernmental cooperation. In contrast, to achieve the EU's GHG emission target, the NECPs only specify the measures through which the national emission targets in non-ETS sectors will be implemented and achieved, whilst national contributions are still determined at EU level by the Effort Sharing Regulation (EP and CoEU 2018a, Annex I).

However, even for renewables and energy efficiency, aspects of the target setting process suggest that it is not strictly bottom-up. The plans, whose framework is outlined in Annex 1 of the regulation, are based on templates produced by the Commission.[7] The EP successfully pushed for a transparent process for producing national plans (EP 2018), through their publication and public consultations involving national parliaments, local and regional authorities, as well as civil society (EP and CoEU 2018b, art. 10 & 11). They will be produced in a structured, transparent, iterative process between the Commission and Member States (EP and CoEU 2018b, article 1). Whilst member states supported intermittent reports on their national climate and energy plans, the EP wanted regular and more comprehensive reporting (interviews 7 and 8). The negotiated outcome was a close oversight role for the Commission in the creation of national plans, with a role in reviewing biennial progress reports on their implementation from 2021, to facilitate EU level aggregation and assessment (EP and CoEU 2018b, article 17). Whilst framed as 'Better Regulation', with streamlined and minimised reporting obligations (interview 5), a number of member states perceive that their preferences have been overruled and that: a) the information reporting obligations are the same or have increased (interview 13); and b) that these obligations grant considerable authority to the Commission to monitor member state policy planning and implementation, and to interpret its role (interviews 7, 9, 11, 13, 16):

"[I]t is not a renationalisation of energy policy, totally the contrary. New governance is what the Commission wants. It is high on the agenda now, and it could continue to be so – and be political. Or it could be low on the agenda and technical … There is an option to really put pressure on national politics and national ministries. There is total discretion legally" (interview 14).

The Commission's first review of draft NECPs at the end of 2018 found that whilst one third of member states were judged to have submitted (sufficiently) ambitious contributions to the EU's renewable energy target, there was an overall gap in ambition. As a result, the Commission recommended that several member states 'reconsider their level of ambition' ahead of final submissions, 'increasing national contributions as appropriate' (European Commission 2019, 3). The same Commission review finds a 'substantial gap' for energy efficiency, with only a few member states proposing 'a sufficient level of contributions for 2030' (European Commission 2019, 4), and that: 'all Member States whose contributions are assessed as not sufficient at this stage are recommended to review them and consider increasing the level of ambition' (European Commission 2019, 5).

In 2023, the Commission will review progress towards the headline 2030 EU targets and achieving its commitments as part of the Paris Agreement (EP and CoEU 2018a, art. 14). It will assess if there is an 'ambition gap' and/or an 'implementation gap'. This review date is considered 'an open door to go higher' (interview 12), and 'has the explicit mention that the targets can only be reviewed upwards, so it is a kind of a ratcheting-up principle' (interview 15). This refers to the clause in both the amended energy efficiency and recast renewable energy Directives (EP and CoEU 2018c, 2018d, article 3).

Compliance mechanisms: peer review and the shadow of the community method

Without binding national targets for renewables and energy efficiency, there is a more limited threat of infringement proceedings from the CJEU. Member states are only obliged: to create national integrated climate and energy plans, and where relevant to respond to Commission recommendations issued on renewables (and possibly other Energy Union objectives) when the EU is collectively adjudged to be failing without any strict obligation to implement them; and to address implementation gaps, choosing the appropriate instruments and measures (EP and CoEU 2018b, article 31). The Commission can initiate infringement proceedings for incorrect or partial implementation (for instance incomplete national plans), delays, or failure to attempt to address the implementation gap, but not for failure to actually achieve renewable and energy efficiency targets. However, the Governance Regulation provides the Commission with more power to consult with member states on plans (interview 8), and to monitor them.

Peer pressure is often identified as a central tool to ensure ambition and compliance. Greater transparency will allow civil society organisations at the EU and national levels to track progress and identify 'leaders' and 'laggards' (interview 10). The binding template is considered a: ' … tool to shape the political process through obligation of member states to make transparent their contributions and underlying assumptions' (interview 9; also interview 16). The EP strongly advocated increased transparency in the governance mechanism to compensate for relatively soft governance as it 'creates a different level of pressure' (interview 10; also 15). In addition, the EP proposed the inclusion of a formula for the calculation of indicative national renewable contributions, which was supported by the most ambitious member states (interviews 7, 11, 16) and included in the final legislation (EP and CoEU 2018b, Annex II).[8] This formula does not mean a return to binding targets but sets expectations. In the words of an energy official in the Commission: ' … the formula is an assessment tool, it is informal, we will use it if there is a gap. We already tell them [MS] what the minimum requirements are' (interview 10). In fact, in its recommendations of June 2019 on the draft NECPs, the Commission has not shied from using the formula to recommend more ambition on renewable energy from 12 states (Euractiv 2019).

In theory, member states will want to avoid being in the group of countries who are considered as not being ambitious enough and asked to do more (interview 16). This is a key aspect of the regulation: 'at the core is soft power. Naming and shaming' (interview 12). Member states themselves will be able to monitor each other's efforts towards the common objectives. As a senior Commission official summarised: 'Deadlines, dialogue, benchmarking, tracking, monitoring are the key components [providing] teeth' to ensure compliance and to steer an upwards dynamic in terms of implementation and ambition (interview 2; also 5, 6, 9).

A State of the Energy Union report will continue to be produced, now on an annual rather than biannual/ad hoc basis (EP and CoEU 2018b, art., 35). If the EU is not on target to meet its goals, then the Commission will make non-binding recommendations.[9] Whilst the target for renewables is binding only at the EU level, member states are obliged to set out a 'mandatory baseline share' and national indicative trajectory from 2021 to 2030. This is a bottom-up mechanism of setting targets, but if renewables are below this then the Commission will ask for additional measures from Member States, who *should* then close this gap. Yet, due to national sensitivities, if the national gap is not closed the Commission is clear that it does not intend to use the reporting and monitoring system to design a system that is 'binding by the back door' (interview 2). Still, member states are obliged to take 'due account' of these recommendations and also provide and make public reasons for not addressing them 'in a spirit of solidarity between Member States and the Union and between Member States' (EP and CoEU 2018b, article 34).[10] Ultimately, the recommendations are considered 'political': 'It is peer pressure. Member states will expose themselves to their peers if they don't do their part' (interview 10; also interview 11).

Beyond reporting, monitoring and peer pressure the 'shadow of hierarchy' is retained. Some parts of related legislation are legally binding, such as sub-targets and objectives enshrined in the Energy Performance in Buildings Directive, the Energy Efficiency Directive or eco-design directives (Dupont 2020). These provisions could be legally enforced to help close the implementation gap. Moreover, if a delivery gap emerges at the EU rather than national level, the regulation explicitly states that the Commission 'shall propose measures and exercise its powers at Union level in order to ensure the collective achievement of those objectives and targets' (EP and CoEU 2018b, article 31). National measures should be prioritised but amending related EU legislation (on renewable energy, energy efficiency, or ecodesign for instance) to strengthen the targets, obligations and monitoring process remain on the table, creating an additional pressure for laggards to comply (interviews 11, 15 and 16). This option would require the agreement of the co-legislators and would therefore necessitate a shift in the position of key member states. This is potentially achievable if the Governance Regulation proves to be too weak to meet the Council's stated ambitions, in a context where decarbonisation rises up the national and EU agendas, and the Commission is tasked with proposing alternatives.

Discussion

Institutional power balance: authority contestation and institutional compromise

As part of flagship Energy Union project, the case of the Governance Regulation reveals the ongoing authority contestation between member states and supranational institutions. Early on, member states took a strong stance to preserve their guarded sovereignty in defining their energy mix and policies and avoid any solution 'imposed by Europe' (interview 2). The development of a formal governance framework enshrined into legislation is therefore, from the perspective of the Commission and European Parliament, only a second-best solution. It is a functionally driven move, motivated by the need to meet the EU's climate commitment within the Paris framework, in the absence of binding national targets for renewable energy and energy efficiency, and while respecting member states' sovereignty concerns.

The Governance Regulation heralds a bi-directional authority shifts in climate and energy governance (see Herranz-Surrallés, Solorio, and Fairbrass 2020). On the one hand there is a downward shift towards a partial reclaiming of authority compared to the 2020 framework, at least concerning the definition and enforcement of national renewable energy objectives, with a reduced role for the Court.[11] At the same time, there is also an upward authority shift towards the EU level in terms of planning and compliance monitoring. NECPs are produced and implemented by member states under the Commission's guidance, covering mainly climate and clean energy objectives, but also other aspects related to the internal market. This suggests that the changes do not conform to a strict pattern of re-nationalisation or disintegration.

The Governance Regulation creates a process of 'harder soft governance' (Ringel and Knodt 2018) – partially inspired by the European Semester system[12] – to manage authority conflicts in energy policy, with formal adjudication restricted to certain aspects (see Herranz-Surrallés, Solorio, and Fairbrass 2020). Yet, there is also a limited shift in authority to the Commission as an agent empowered to monitor and advise member states on national plans and strategies, including through the use of the indicative renewable energy formula. There is also a potential further shift to the Commission in terms of recommendations and additional measures in case of failure, if the change in the governance process, and greater self-governance, is in danger of failing to achieve binding EU targets. In the case where there is a delivery gap between headline goals set by the member states themselves and their implementation as part of the Governance Regulation, it is likely that substance-based contestation over the level of ambitions and best way to achieve them will become explicit, with conflicts emerging between leaders and laggards. In turn this is likely to reopen discussions and contestation on sovereignty delegation if the relatively informal or softer authority conferred to supranational institutions in the Regulation has been insufficient to prevent non-compliance and free riding.

Unlike economic governance within the Semester, there will be no binding country specific recommendations from the Commission. The Commission has the power to use legal sanctions in the form of infringement procedures only for non-submission, delays or improper realisation of national plans and not for any lack of ambition within them or failure to comply with Commission recommendations. However, member states with an initial ambition gap, or later an implementation gap, will be under domestic and peer member state pressure to address this failure, the Commission using bottom-up rather than the top-down pressure used previously. Politicisation is therefore embedded in the governance, to compensate for the softer judicial pressure. As a national representative argued, the process is designed ' ... to bring politics into energy policy' (interview 14). Political contestation in implementation is expected to compensate for the relative depoliticisation of target setting in decision-making resulting from the loss of national renewable targets.

Secondly, the shadow of hard governance through the traditional Community method remains. This includes potential new legislative proposals to address collective EU failure. Arguably, member states are far from united and a blatant failure to keep on track towards EU objectives could lead to changing Council positions and apossible shift of authority to supranational institutions accompanied by 'harder' legislation. A partial move away from supranational governance is made contingent on member states achieving the commitments (emissions, renewables and energy efficiency) they have made in the European Council and as part of the Paris Agreement. Oberthür (2019) concludes that on balance the 2030 governance framework is no less stringent than previously.

The Energy Union governance framework as embedded intergovernmentalism

Whilst most aspects of the clean energy package remain within the scope of the Community method, the process has been dominated by governments from the agenda setting phase to the negotiations. As part of the European Council, member states set the framework and the headline goals within which the Commission has had to work, notably at the October 2014 summit. Yet, a careful tracing of the negotiations process suggests that supranational institutions have not been merely reactive and have contributed to shape the process and outcome, aided by divisions and shifts among member states themselves.

If the Commission refrained from directly challenging the Heads of State and Governments when drafting legislative proposals, as expected by NI, it also promoted a stronger role for itself in monitoring national plans and progress towards the EU objectives. As for the EP it successfully pushed for more ambitions (if not for more binding objectives), aided by a shift in the power balance in the Council in the last phase of the negotiations. The EP was also instrumental in strengthening the role of the Commission in monitoring progress and compliance. The compromise reflects a balancing act in the Council as well, with some governments having concerns about free riding by less ambitious member states while others were worried by the potential discretion of the Commission in assessing national plans and objectives.

Energy policy does not fit neatly into the new intergovernmentalist framework. Member states are indeed central, and their collective authority (and relations) key to explain policy outcomes. Yet, in line with critics of new intergovernmentalism, we see cooperation and conflict as shaped by the EU institutional environment. The distribution of authority is far from clearly demarcated and stable, with potential for both informal competence creep and renationalisation. The outcome of the Governance Regulation negotiations is a framework in which the governments remain in the driving seat when defining their national headline objectives and the policy instruments to achieve them but are embedded in a process of monitoring and peer review in which the Commission has a key political role to play. It has authority to guide member state policy plans and facilitate compliance, with some discretion in how it applies the Regulation's principles. In terms of institutional power balance, new intergovernmentalism places authority on the most sovereignty sensitive issues firmly with the Council and European Council, where member states are represented, and consensus achieved. In contrast, in areas where the Community method dominates, authority is distributed among legislative institutions and the involvement of the European Council limited and intermittent. The Governance Regulation does not map neatly on either the traditional Community method or new intergovernmentalism (see Table 1). It is better characterised as a 'hybrid' at the intersection between intergovernmental governance and the Community method, as a form of embedded intergovernmentalism.

Conclusion

Firstly, the Energy Union governance framework illustrates the challenge of addressing enduring sovereignty concerns of member states and providing them with a degree of autonomy regarding their policy choices, while ensuring that headline targets and objectives collectively agreed are met. There has been a high level of contestation, between member states and supranational institutions, but also among member states themselves, regarding the desirable pattern of authority distribution, resulting in a bi-directional authority shift. Upward delegation of authority was to a significant extent successfully resisted by the member states on sovereignty grounds, and there has been a partial renationalisation of (renewable) energy policy. Yet, this intergovernmental shift is only partial and embedded, as there has also been an upward authority shift in order to mitigate concerns regarding the efficacy of an increasingly bottom-up process. The Commission is empowered to monitor, publicise and guide member states' policies. Ultimately, the Energy Union governance framework attempts to reconcile the objectives of balancing sovereignty with solidarity and flexibility without free riding; the text of the regulation mentions in its preamble (para. 59) the need 'to avoid the "free rider" effect' as a motivation for a close monitoring of member states' national renewable energy trajectory.

Secondly, the regular process of monitoring, reporting and revising is intended to provide flexibility in terms of targets and policy setting in a rapidly changing technological and political environment. Rather than a 10-year plan working towards fixed targets for 2030, as was the case for the EU's 2020 climate and energy framework, the Governance

Regulation provides for multiple points for stocktaking and corrective action if members states are collectively under delivering. This iterative process also offers opportunities to react to potential technological and political changes – increasing public pressure, falling technology costs, the results of international climate negotiations – and integral to its design is facilitating a ratcheting up of climate ambition.

Notes

1. Security, solidarity and trust; a fully integrated energy market; energy efficiency; decarbonisation; and research, innovation and competitiveness.
2. The EU wide binding GHG targets rests on binding national targets in sectors not covered by the EU Emission Trading Scheme (ETS) as part of the Effort Sharing Regulation, and obligations which are binding on companies in the sectors covered by the ETS.
3. In June 2018, the German Economy and Energy minister effectively vetoed any higher ambition than 32% for renewables, setting an upper limit in negotiations (Simon 2018b).
4. The Council proposal (2022-16%, 2025-40%, 2027-60%) compared to the EP proposal (2022-20%, 2025-45%, 2027-70%) (Council of the EU 2018). 18, 43 and 65 was eventually agreed (EP and CoEU 2018a, art. 4.2).
5. The EP proposed that long-term emission strategies should run to 2100, and crucially that by 2050 they should show zero emissions, with negative emissions after (EP 2018).
6. The net zero commitment was eventually endorsed by the European Council at its 12 December 2019 summit, although at the price of constructive ambiguity: the conclusions state that 'one member state [Poland], at this stage, cannot commit to implement this objective' (European Council 2019, 1).
7. While the Commission proposed that these templates *shall* (mandatory) be used, it was eventually watered down to *should* (advisory) (EP and CoEU 2018b).
8. Germany, Spain and Sweden were amongst the member states pushing for a binding formula initially, though this was abandoned in 2014 (interviews 11 and 16).
9. Sweden and Luxembourg were amongst the minority willing to have binding recommendations (interview 8).
10. This was weakened from the Commission's position of wanting member states to 'take utmost account' of these (European Commission 2016).
11. In contrast, the climate targets defined in the Effort Sharing Regulation remain binding and potentially enforceable through court proceedings, while the headline energy efficiency target is still only indicative.
12. As one interviewee explained, the Governance Regulation reflects the Council's view that 'the whole governance of the Climate should follow the model of the European Semester' (interview 15).

Acknowledgments

We thank the editors of the Special Issue, Anna Herranz, Jenny Fairbrass and Israel Solorio, for their insightful comments, hard work and patience in bringing this collective project to completion. We are also grateful to Katja Biedenkopf, Quentin Génard and the anonymous reviewers for their comments and suggestions. All remaining errors remain of course entirely our own.

Interviews

Interview 1: Member state official, Brussels, 09.04.18
Interview 2: Senior Commission official, Brussels, 10.04.18
Interview 3: Electricity industry representative, Brussels, 10.04.18

Interview 4: Senior gas industry representative, Brussels, 11.04.18
Interview 5: Commission official, Brussels, 13.04.18
Interview 6: Think tank official, Brussels, 13.04.18
Interview 7: Member state official, phone, 17.04.18
Interview 8: Member state official, Brussels, 29.06.18
Interview 9: DG Energy official, Brussels, 02.07.18
Interview 10: Commission official, Brussels 02.07.18
Interview 11: Member state official, Brussels, 02.07.18
Interview 12: Journalist, Brussels, 04.07.18
Interview 13: Member state official, Brussels, 04.07.18
Interview 14: Member state official, Brussels, 05.07.18
Interview 15: MEP's assistant, Brussels, 05.07.18
Interview 16: Member state official, Brussels, 05.07.18
Interview 17: Journalist, Brussels, 06.07.18
Interview 18: MEP, email, 13.07.18

Disclosure statement

No potential conflict of interest was reported by the authors.

ORCID

Tomas Maltby (iD) http://orcid.org/0000-0002-8434-9749

References

Alexandrova, P., and A. Timmermans. 2015. "Agenda Dynamics on Energy Policy in the European Council." In *Energy Policy Making in the EU*, edited by J. Tosun, S. Biesenbender and K. Schulze, 41–61. London: Springer.

Andoura, S., L. Hancher, M. Van der Woude, and J. Delors. 2010. *Towards A European Energy Community: A Policy Proposal*. Brussels: Notre Europe.

Bickerton, C. J. 2012. *European Integration: From Nation-states to Member States*. Oxford: Oxford University Press.

Bickerton, C. J., D. Hodson, and U. Puetter. 2015b. "The New Intergovernmentalism and the Study of European Integration." In *New Intergovernmentalism: States and Supranational Actors in the Post-Maastricht Era*, edited by C. J. Bickerton, D. Hodson, and U. Puetter, 1–48. Oxford: Oxford University Press.

Bickerton, C. J., D. Hodson, and U. Puetter. 2015a. "The New Intergovernmentalism: European Integration in the post-Maastricht Era." *Journal of Common Market Studies* 53 (4): 703–722.

Bocquillon, P., and M. Dobbels. 2014. "An Elephant on the 13th Floor of the Berlaymont? European Council and Commission Relations in Legislative Agenda Setting." *Journal of European Public Policy* 21 (1): 20–38. doi:10.1080/13501763.2013.834548.

Buchan, D. 2009. *Energy and Climate Change: Europe at the Crossroads*. Oxford: Oxford University Press.

Bulmer, S. 2015. "Understanding the New Intergovernmentalism." In *The New Intergovernmentalism: States and Supranational Actors in the Post-Maastricht Era*, edited by C. J. Bickerton, D. Hodson, and U. Puetter, 289–303. Oxford: Oxford University Press.

Bürgin, A. 2015. "National Binding Renewable Energy Targets for 2020, but Not for 2030 Anymore: Why the European Commission Developed from a Supporter to a Brakeman." *Journal of European Public Policy* 22 (5): 690–707. doi:10.1080/13501763.2014.984747.

CoEU (Council of the EU). 2018. "Preparation of Second Trilogue." *General Secretariat of the Council, 6794/18*, April 5.

Dehousse, R., ed. 2011. *The 'community Method': Obstinate or Obsolete?* New York: Palgrave.

Dupont, C. forthcoming. "Defusing Contested Authority: EU Energy Efficiency Policymaking." *Journal of European Integration* 42 (1).

Eberlein, B. 2008. "The Making of the European Energy Market: The Interplay of Governance and Government." *Journal of Public Policy* 28 (1): 73–92. doi:10.1017/S0143814X08000780.

Eising, R. 2002. "Policy Learning in Embedded Negotiations: Explaining EU Electricity Liberalization." *International Organization* 56 (1): 85–120. doi:10.1162/002081802753485142.

EP (European Parliament). 2018. "Governance of the Energy Union." *P8_TA(2018)0011*, January 17.

EP and CoEU. 2018a. "REGULATION (EU) 2018/842 on binding annual greenhouse gas emission reductions [...]." June 19.

EP and CoEU. 2018b. "REGULATION (EU) 2018/1999 On the Governance of the Energy Union and Climate Action." December 21.

EP and CoEU. 2018c. "DIRECTIVE (EU) 2018/2001 ... on the promotion of the use of energy from renewable sources (recast)." December 21.

EP and CoEU. 2018d. "DIRECTIVE (EU) 2018/2002 amending Directive 2012/27/EU on energy efficiency." December 21.

Euractiv. 2019. "NECP Recommendations." Accessed 14 June 2019. https://www.euractiv.com/wpcontent/uploads/sites/2/2019/06/NECP-recommendations.pdf

European Commission. 2015. "A Framework Strategy for A Resilient Energy Union with A Forward-Looking Climate Change Policy." *COM(2015) 80 final*, February 25.

European Commission. 2016. "Commission Staff Working Document Fitness Check: Reporting, Planning and Monitoring Obligations in the EU Energy Acquis." *SWD(2016) 397 final*, November 30.

European Commission. 2019. "United in Delivering the Energy Union and Climate Action - Setting the Foundations for a Successful Clean Energy Transition." *COM(2019) 285 final*, June 18.

European Council. 2014. "European Council (23 and 24 October 2014) – Conclusions." *EUCO 169/14*, October 24.

European Council. 2015. "European Council Meeting (19 and 20 March 2015) – Conclusions." *EUCO 11/15*, March 20.

European Council. 2018. "European Council Conclusions on Jobs, Growth and Competitiveness, as Well as Some of the Other Items (Paris Agreement and Digital Europe)." *EUCO 1/18*, March 23.

European Council. 2019. "European Council Meeting (12 December2019) – Conclusions." *EUCO 12/12/19*, December 12.

Herranz-Surrallés, A., I. Solorio, and J. Fairbrass. forthcoming. "Renegotiation Authority in the Energy Union: A Framework for Analysis." *Journal of European Integration* 42 (1).

Hodson, D. 2013. "The Little Engine that Wouldn't: Supranational Entrepreneurship and the Barroso Commission." *Journal of European Integration* 35 (3): 301–314. doi:10.1080/07036337.2013.774779.

Hodson, D., and U. Puetter. 2019. "The European Union in Disequilibrium: New Intergovernmentalism, Postfunctionalism and Integration Theory in the post-Maastricht Period." *Journal of European Public Policy* 26 (8): 1153–1171. doi:10.1080/13501763.2019.1569712.

Hooghe, L., and G. Marks. 2009. "A Postfunctionalist Theory of European Integration: From Permissive Consensus to Constraining Dissensus." *British Journal of Political Science* 39 (1): 1–23. doi:10.1017/S0007123408000409.

Judge, A., and T. Maltby. 2017. "European Energy Union? Caught between Securitisation and 'riskification'." *European Journal of International Security* 2 (2): 179–202. doi:10.1017/eis.2017.3.

Maricut, A., and U. Puetter. 2018. "Deciding on the European Semester: The European Council, the Council and the Enduring Asymmetry between Economic and Social Policy Issues." *Journal of European Public Policy* 25 (2): 193–211. doi:10.1080/13501763.2017.1363271.

Matlary, J. H. 1996. *Energy Policy in the European Union*. Berlin: Springer.

Morata, F., and I. Solorio, eds. 2012. *European Energy Policy: An Environmental Approach*. Cheltenham: Edward Elgar.

Nelsen, A. (2014). "Denmark Signals Fight for Tougher 2030 Climate and Clean Energy Goals." *Euractiv*, January 27.

Novak, S. 2013. "The Silence of Ministers: Consensus and Blame Avoidance in the Council of the European Union." *Journal of Common Market Studies* 51 (6): 1091–1107. doi:10.1111/jcms.12063.

Oberthür, S. 2019. "Hard or Soft Governance? the EU's Climate and Energy Policy Framework for 2030." *Politics and Governance* 7 (1): 17–27. doi:10.17645/pag.v7i1.1796.

Piris, J. C. 2010. *The Lisbon Treaty: A Legal and Political Analysis*. Cambridge: Cambridge University Press.

Puetter, U. 2012. "Europe's Deliberative Intergovernmentalism: The Role of the Council and European Council in EU Economic Governance." *Journal of European Public Policy* 19 (2): 161–178. doi:10.1080/13501763.2011.609743.

Puetter, U. 2014. *The European Council and the Council: New Intergovernmentalism and Institutional Change*. Oxford: Oxford University Press.

Ringel, M., and M. Knodt. 2018. "The Governance of the European Energy Union: Efficiency, Effectiveness and Acceptance of the Winter Package 2016." *Energy Policy* 112: 209–220. doi:10.1016/j.enpol.2017.09.047.

Schimmelfennig, F. 2015. "What's the News in 'new Intergovernmentalism'? A Critique of Bickerton, Hodson and Puetter." *Journal of Common Market Studies* 53 (4): 723–730. doi:10.1111/jcms.12234.

Schmidt, S. K. 1998. "Commission Activism: Subsuming Telecommunications and Electricity under European Competition Law." *Journal of European Public Policy* 5 (1): 169–184. doi:10.1080/13501768880000081.

Schmidt, V. A. 2018. "Rethinking EU Governance: From 'old' to 'new' Approaches to Who Steers Integration." *Journal of Common Market Studies* 56 (7): 1544–1561. doi:10.1111/jcms.12783.

Simon, F. 2018a. "Parliament Backs 'net-zero' Carbon Emissions by 2050." *Euractiv*, 18 January.

Simon, F. 2018b. "Germany Pours Cold Water on EU's Clean Energy Ambitions." *Euractiv*, June 11.

Slominski, P. 2016. "Energy and Climate Policy: Does the Competitiveness Narrative Prevail in Times of Crisis?" *Journal of European Integration* 38 (3): 343–357. doi:10.1080/07036337.2016.1140759.

Solorio, I., and H. Jörgens. forthcoming. "Contested Energy Transition? Europeanization and Authority Turns in EU Renewable Energy Policy." *Journal of European Integration* 42 (1).

Stone Sweet, A., and W. Sandholtz. 1997. "European Integration and Supranational Governance." *Journal of European Public Policy* 4 (3): 297–317. doi:10.1080/13501769780000011.

Szulecki, K., S. Fischer, A. T. Gullberg, and O. Sartor. 2016. "Shaping the 'energy Union': Between National Positions and Governance Innovation in EU Energy and Climate Policy." *Climate Policy* 16 (5): 548–567. doi:10.1080/14693062.2015.1135100.

Thaler, P. 2016. "The European Commission and the European Council: Coordinated Agenda Setting in European Energy Policy." *Journal of European Integration* 38 (5): 571–585. doi:10.1080/07036337.2016.1178252.

van Renssen, S. 2014. "The EU's Great 2030 Energy and Climate Compromise." *Energy Post*, October 24.

Vandendriessche, M., A. Saz-Carranza, and J. M. Glachant 2017. "The Governance of the EU's Energy Union: Bridging the Gap?" *EUI Working Paper 2017/51*. San Domenico di Fiesole: Robert Schuman Centre for Advanced Studies.

Wilson, A. B. 2018. "*Briefing: Governance of the Energy Union*." *European Parliamentary Research Service*, 16 March.

Private authority in tackling cross-border issues. The hidden path of integrating European energy markets

Sandra Eckert and Burkard Eberlein

ABSTRACT
We investigate private authority in European Union (EU) energy governance in order to address two research questions: First, how has authority been conferred on, and acquired by private actors? Second, to what extent has this lateral shift of authority been contested and on which grounds? The paper links the literatures on regulatory governance and private authority. This allows us to shed light on an issue that tends to be neglected in the discussion about the transfer of competencies in the energy field: the horizontal transfer of authority. In our case study about the role of transmission system operators (TSOs) in the creation of an internal electricity market, we identify three distinct settings where both the level of sovereignty-based contestation and the shift towards private authority vary. We find that private rulemaking has gained in importance due to functional expertise requirements, but also because it provides an escape route in a context of political contestation.

Introduction

From mid-January to March 2018 electric clocks slowed across 25 European countries. This was caused by a decrease in the electric frequency in the continental transmission network. The imbalance in the European power system originated in Serbia and Kosovo: Serbia prevented Kosovo from importing energy from neighbouring Albania, while Serbia's electrical power grid company Elektromreža Srbije blamed Kosovo for withdrawing uncontracted electric energy from the synchronised European grid (EurActiv 2018). The European association of transmission system operators (TSOs), ENTSO-E, reacted swiftly to find a technical solution. Nevertheless, the power deviations originating from the Serbian control area persisted throughout 2018. In a press release published in November 2018, ENTSO-E stated that in 'the general public interest' its 'efforts are aimed at ending further time deviations and financial implications and ultimately avoid a potential black-out in Europe' (ENTSO-E 2018).

The above example demonstrates how electricity grid operators play a crucial role as guarantors of secure and reliable operation of an interconnected grid. Moreover, they have a mandate in realising two key objectives of European energy policy as stated in the Lisbon Treaty (article 194 TFEU): '(a) ensure the functioning of the energy market' and '(d) promote the interconnection of energy networks'. In the new context of integrated markets, their long-standing role of providing cross-border capacity for the flow of electricity (and gas) gives these organisations new power positions as guarantors of both trade flows and of security of electricity supply. Drawing on their legitimacy as 'connectors' in the EU electricity market, network operators have been able to shape the emerging European policy framework and acquire a regulatory role formalised by European legislation. To better understand this development in European energy governance, and discuss its effects, we address two research questions: *First: to what extent has authority been conferred on, and acquired by private actors? Second: have lateral shifts of authority, where they occurred, been subject to contestation?*

Infrastructure-related aspects merit our attention (Crisan and Kuhn 2016), since this dimension has been understudied in the analysis of energy market integration (Labelle 2016, 154) and remains under-researched in the literature on energy governance. Authority in European energy policy has not only moved upwards towards supranational policy-makers, but has also been displaced laterally towards private actors (Kahler and Lake 2004; Eckert 2011; Herranz-Surrallés, Solorio, and Fairbrass forthcoming). This is especially salient seen that transmission system operators (TSOs) provide a public good – security of supply and secure operation of the grid – funded by taxpayers' money. Whether network operators hold authority is thus not only a matter of competence and regulatory power, but also one relating to the adequate cost and provision of a publicly funded infrastructure. Ultimately, the cost of infrastructure expansion is passed on to consumers.

In our contribution to this Special Issue, we pay specific attention to the lateral shift of authority towards infrastructure operators. We first develop a conceptual framework and derive hypotheses from the literature on private authority and regulatory governance. The case study proceeds in two steps: in a first step, we examine how authority has evolved over time, focusing on the role of TSOs and infrastructure-related cross-border cooperation. In a second step, we conduct a case study with three constellations showing the intensity and direction of authority displacement. Finally, we discuss whether and why such authority displacement has been contested.

Displacing authority laterally

Ever since Majone published his seminal work on the rise of the European regulatory state (Majone 1996), there has been a rapidly growing literature on regulatory policies and governance in EU scholarship (Lodge 2008; Eckert 2011). Especially the role of so-called non-majoritarian institutions has been at the center of scholarly interest (Thatcher and Sweet 2002). By contrast, the regulatory role of private actors has not received the same level of attention. There is a thriving literature on private governance (Knill 2001) and industry self-regulation, with a special focus on environmental policy (Héritier and Eckert 2008; Rottmann and Lenschow 2008). Yet, the discussion about energy policy so far has not paid adequate attention to the role of private actors. With the focus on upward shifts of authority from Member States to the EU, the degree of centralised administrative capacity tends to be overstated. By contrast, the lateral displacement of authority, from public to private actors, has been largely overlooked. Such lateral shifts will likely lead to a horizontal conflict of authority in which the balance between public and market actors requires recalibration (Kahler and Lake 2004). To fully grasp not only how and why public actors confer power on private actors, but also how private actors acquire authority over time, we suggest to consider both the demand (public) and supply (private) side of the equation drawing on resource dependence theory (Pfeffer and Salancik 2003, xiii). The analysis of the power base of both sides, depending on their respective resources, makes it possible to formulate expectations about the type of exchange in which public and private actors engage.

The demand side: the need for expertise and operational capacity

The literature on private authority (Cutler, Haufler, and Porter 1999; Green 2014) has discussed extensively why and how public actors confer authority on private actors. In their seminal work on private authority, Cutler et al. posit that authority emerges '[...] when an individual or organisation has decision-making power over a particular issue area and is regarded to exercise that power legitimately' (Cutler, Haufler, and Porter 1999, 5). Private authority is thus a special case of legitimate power held by private actors. The term private authority is thus defined in delineation from public authoriy and the state mono-poly of decision-making power and, for instance, not with regards to public or private ownership. It is in this sense that we speak of private actors in this paper. Private actors are typically the addressees of regulation, and only where they exercise rule-making tasks do they hold private authority.

As a primary source of authority, the literature singles out expertise (Cutler, Haufler, and Porter 1999, 4; Green 2014, 7). Some have suggested the term 'epistemic authority' where authority draws on expertise and competence (Sending 2015; Quack 2016). Typically, operators of critical infrastructures derive their authority from a legacy in providing a public good (Cutler, Haufler, and Porter 1999, 4). In this special case authority therefore draws on the operators' legacy as a high reliability organisation (Eckert 2019, 185–221), which 'must not make serious errors because their work is too important and the effects of their failures too disastrous' (Consolini and LaPorte 1991, 19). The para-mount goal, therefore, is to ensure 'failure-free organizational performance' (Consolini and LaPorte 1991, 20). When public actors confer authority on private actors, they thus seek to secure the required level of expertise and operational capacity. In exchange, they can offer to legitimise private governance, either *ex post* or *ex ante*. Here, Green's distinction between delegated and entrepreneurial private authority provides a helpful analytical tool (Green 2014, 33–6). The difference between these types of authority relates to the origins of authority: 'Claims of delegated authority are ultimately derived from the state' (Green 2014, 7). According to Green, delegated authority is *de jure* and involves a principal–agent relation between public and private actors. It is more likely to occur where powerful states are in agreement with each other, and where a strong focal institution that can potentially monitor private action if necessary exists (Green 2014, 17, 41). By contrast, entrepreneurial authority 'does not originate with the state' (Green 2014, 7). Rather, it results from 'governance failure' (Green 2014, 17). Given that 'any private actor that projects authority, must persuade others to adopt its rules or practices'

(Green 2014, 7), entrepreneurial authority only occurs where private actors succeed in this endeavour. As private actors cannot rely on coercion as a source for gaining legitimacy, they have to mobilise other sources such as expertise or moral arguments. Green goes on to argue that, in contrast to delegated authority, 'entrepreneurial authority is *de facto*' (Green 2014, 7). The distinction between these types of authority also involves a temporal distinction when it comes to governmental consent: 'the governed grant their consent *ex ante* in the case of delegated authority, whereas it tends to be *ex post* in entrepreneurial authority' (Green 2014, 7).

The supply side: acquiring authority and preventing onerous regulation

The differentiation between delegated and entrepreneurial authority emphasises the proactive role of both public and private actors. Or, to use the wording of Cutler et al.: 'the cooperation among firms is either given legitimacy by governments or legitimacy is acquired through the special expertise or historical role of the private sector participants.' (Cutler, Haufler, and Porter 1999, 4). Besides the reputational gains that private actors can derive from entrepreneurial authority, it is the prevention of onerous regulation that motivates them to become proactive. In accordance with the 'regulatory threat' scenario the mere possibility of public intervention alters incentive structures for private actors (Halfteck 2008; Glachant 2003; Halfteck 2006). Especially in heavily regulated sectors such as energy, the uncertainty that infrastructure providers face lies in the regulatory risk of onerous and costly intervention by rulemakers. If they can obtain delegation – that is, private authority – they can contain this risk. The primary external resource they depend on is regulatory approval and consent, whereas they offer expertise and assets in exchange.

Conceptualising authority shifts in times of contestation

Our contribution highlights the general importance of private authority in European energy governance. More specifically, we argue that private authority is not necessarily less likely to materialise in times of politicisation and contestation (Herranz-Surrallés, Solorio, and Fairbrass forthcoming; Tosun and Mišić forthcoming). This claim may appear counter-intuitive. Normally, one would expect that political actors seek to (re)gain control of private actors when issues become politically salient or contested – unless, of course, political actors use delegation to private actors to shift blame (Hood 2002). In addition, given a generally negative public attitude towards both business power and EU involvement (Fuchs, Gumbert, and Schlipphak 2017, 318), one would expect EU policymakers to want to be seen taking action that goes against big business interests. However, technical issues concerning infrastructure operation often only attract very low levels of direct public attention. It follows that the political benefits of exercising control may be limited.

In order to systematically capture various settings which may result from authority shifts we focus on the interlinkage between a) shifts towards private authority and b) the degree to which the increase of supranational competenies is contested. Depending on the character of shifts towards private authority and of the level of contestation of supranational competencies (strong or weak), this results in one of four different constellations: (1) the *Escape Route* (strong/strong); (2) the *Political Route* (weak/strong); (3) the *Technocratic Route* (weak/weak); and (4) the *Technical Route* (strong/weak).

Table 1. Conceptualising shifts towards private authority.

Authority Shift	Strong	Weak
Sovereignty-based Contestation		
Strong	(1) *Escape Route* delegated private authority	(2) *Political Route* political control
Weak	(3) *Technical Route* entrepreneurial private authority	(4) *Technocratic Route* administrative authority

Source: own table.

We posit that sovereignty-based contestation (discussed by Bocquillon and Maltby forthcoming) targets vertical rather than lateral shifts. As a consequence, shifting competencies towards private actors can constitute an (1) *Escape Route* even in settings with high levels of contestation as depicted in Table 1. Such an escape route is particularly appealing in a setting where political actors face a risk of policy deadlock, yet do not perceive immediate benefits of exerting control. Shifting competencies towards private actors is especially attractive in cases where entrepreneurial authority already existed prior to the attempts to generate European capacity on a certain issue. Here we can expect entrepreneurial authority to not only receive ex post endorsement, but also for policymakers to specify a formalised mandate resulting in delegated authority.

H1 A high level of sovereignty-based contestation combined with a low level of perceived political benefits of exerting control is conducive to a lateral shift of authority towards private actors (Escape Route).

The alternative to this escape route is the (2) *Political Route*, where no such shift to private authority occurs. This would be the type of outcome expected in a highly politicised environment. According to post-functionalist logic, we would expect this to be the predominant outcome in the current stage of integration. However, as politicisation is rooted in sovereignty-based contestation, we could simply see a lateral shift of authority as stated by H1. An additional requirement for the political route to be taken is an immediate benefit of exerting political control. This might occur in a situation where political actors compete over the control and direction of market regulation in order to achieve varying (re)distributive effects. This competition can have either a territorial dimension, opposing national versus European interests; or it can have a horizontal dimension, with diverging interests of market actors (e.g. incumbents versus new entrants; traditional energy sources versus renewables). A political benefit calculation may thus block the path to private actors acquiring authority.

H2 A high level of sovereignty-based contestation combined with a high level of perceived political benefits of exerting control prevents a lateral shift of authority towards private actors (Political Route).

We discern a third, intermediate route where contestation is weak, but the shift towards private authority is strong, namely the (3) *Technical Route*. In this setting, authority shifts towards private actors will go largely unnoticed as there are no perceived political benefits involved. Delegation can be framed as being limited to purely technical matters that are not of wider political relevance. Here it is quite likely that entrepreneurial authority will do the job, unencumbered by much political attention, let alone contestation.

H3 A low level of sovereignty-based contestation combined with a low level of perceived political benefits of exerting control is conducive to a lateral shift of authority towards private actors (Technical Route).

Finally, a setting which combines weak levels of contestation and a weak shift towards private authority is the (4) *Technocratic Route.* In this setting, both political and private actors are largely uninterested in engaging in an exchange of resources. Political actors are not in need of expertise or technical capacity provided by private actors, but can instead delegate a given task to non-majoritarian, technocratic actors. Private actors do either not fear onerous regulatory intervention, possibly because they are well served by the regulatory framework, or they are not affected by the regulation in question.

H4 A low level of sovereignty-based contestation combined with a low level of perceived political and business benefits of exerting control prevents a lateral shift of authority towards private actors (Technocratic Route).

If neither the technocratic route (because of the need to secure expertise and operational capacity), nor the technical route (due to contestation) are available, we expect pro-integrationist, European policy-makers and private actors join forces and engage in some sort of exchange. This will be the case especially in areas where the political route is gridlocked (Héritier 1999). Private actors enjoy some leeway in developing rules and implementing their schemes, while public actors draw on private authority to engage in an incremental path of building up European governance capacity. That is to say that lateral and upward shifts interact, and mutually reinforce each other. More specifically, heightened domestic contestation of upward shifts towards the European level can be to the advantage of private actors, especially where such lateral shifts can be framed as involving purely technical and operational issues. An additional factor facilitating shifts towards private authority is infrastructure operators' strong ties to their national context, while supporting the pro-integrationist agenda The *Escape Route*, therefore, can be considered as an appealing hybrid choice seeking to reconcile supranational and national interests.

While upward shifts of authority will mainly trigger sovereignty-based contestation, lateral shifts will be more about substance-based contestation. Authority shifts towards private actors (technical and escape route) may thus be challenged on substantive grounds. Here, it is other market participants or regulators who will seek to put into question the TSOs' acquired level of private authority by generating and providing counter policy expertise. Private authority becomes subject to the 'politics of expertise' where the various stakeholders involved draw on competing sources of expertise in support of their preferred policy choices (Fischer 1990, 28; Boswell 2009, 5).

In what follows, we will first present the TSOs' role in the evolution of energy governance. Then, we move on to present a case study of cross-border TSO cooperation. We identify three constellations that involve different types of authority displacement. Our analysis draws on document analysis and expert interviews. Thirty semi-structured interviews were conducted with experts at the European Commission DGs (6), at national ministries (3), inside regulators (6), with TSOs (9) and the energy industry (6) between 2005 and 2019.

The private authority of European TSOs

In European energy regulation, infrastructure-related issues have for a long time remained within the remit of national regulatory authorities (Eberlein 2012, 2008), notably because security of supply issues were perceived from a national, rather than a European perspective (Bjørnebye 2006, 335). At the same time, the need for technical cooperation predated the integration of national electricity markets inside the European Communities and the European Union (Eckert 2016). TSOs, as a matter of fact, have engaged in technically driven cooperation for decades, while attempts to build up regulatory oversight at EU-level are a recent phenomenon. It does therefore not come as a surprise that, once policymakers had committed to the market integration agenda, existing private governance structures were used to fill a widening regulatory gap (Eberlein and Grande 2005). This regulatory gap results from governance failure (Green 2014, 17) on the side of public actors. We argue that this authority displacement towards private actors has been largely overlooked in the literature. As a consequence, upward shifts of authority, towards an emerging European administrative space (e.g. Jevnaker 2015), have been overstated as mentioned before. We thus make a significant contribution to a mushrooming literature on energy governance (e.g. from the perspective of experimentalist governance: Mathieu and Rangoni 2019, Rangoni 2019) which so far has unduly emphasised the role of public regulatory agencies and networks.

In our discussion, we proceed as follows: First, we discuss the diversity of TSOs across Europe and their legacy in cooperating on cross-border issues and how this feeds into the European governance architecture. A key point we make is that TSOs have exerted considerable influence on the emerging European governance architecture.

The diversity of TSOs across Europe

TSOs only came into existence as separate entities once vertically integrated companies were compelled to unbundle. In accordance with the different paths of market liberalisation taken across countries, the TSOs in Europe diverge considerably in terms of ownership and unbundling (Meletiou, Cambini, and Masera 2018). Under European law, three types of TSOs are compliant with separation requirements, namely full ownership unbundling (OU), the Independent System Operator (ISO) model, and the Independent Transmission Operator (ITO) model. Under the ISO model, the ownership and operation of the transmission grid must be separated, i.e. the system operator does not dispose of the grid assets, which still belong to an integrated company. Under the ITO model, TSOs may remain part of a vertically integrated undertaking, but have to comply with a minimum degree of organisational separation, operating the network through a subsidiary. All of these three models exist across Europe, while full unbundling and ITO are the two models most frequently used. So far only the UK has fully privatised its network, whereas most countries still have partial or full state ownership. As mentioned before, our conceptualisation of private actors is not based on private ownership, but relates to authority held by actors with a business interest who are typically addressees of regulation. This does apply to European TSOs regardless of their ownership structure.

TSOs incorporate both a European and a national outlook: as owners and operators of infrastructure they have a natural interest in expanding their asset base, for instance by investing in interconnectors (Eyre 2016, 136; Interview COM 2017, 2019b). On the other hand, TSOs very much remain nationally embedded economic agents (also due to their ownership structures), holding a public mandate in a predominantly nationally regulated sector. This is why they maintain close relations with national governments and regulators. The association of TSOs at European level thus embodies both a 'supranational' pro-integrationist stance, but also a 'national' perspective brought in through the members.

The legacy of cross-border cooperation between TSOs

TSOs as the owners of the high-voltage grids to transport electricity, play a central role in the cross-border architecture. As a matter of fact, private authority on cross-border issues predates the Europeanisation of energy policy. Cooperation at regional level started in the early 1950s. In 1951, the Union for the Coordination of Production and Transmission of Electricity (UCPTE) was created in Central Europe. In 1999, resulting from the spin-off of producer interests, UCPTE was transformed into the Union for the Coordination of Transmission of Electricity (UCTE). Other regional bodies followed the example of UCTE, such as NORDEL gathering the Nordic countries in 1963, or the United Kingdom Transmission System Operators Association (UKTSOA) as well as the Association of the Transmission System Operators of Ireland (ATSOI), which were both established in 1999.

These four regional associations initiated the creation of the European Transmission System Operators (ETSO) association, starting to cooperate informally inside EURELECTRIC from 1999 onwards. ETSO was created as a response to an emerging European policy agenda, but its centralised structure was also seen as more appropriate to address cross-border technical issues. The persisting divergence of national and regional practice emerged as a potential threat to secure operation in an increasingly interconnected grid (Interview ETSO 2007). In 2001, ETSO became an international association with direct membership of 32 TSO companies from the 15 countries of the European Union plus Norway and Switzerland. ETSO subsequently enlarged its membership towards the TSOs of East and South-East European countries.

All of the previously mentioned regional sub-organisations and ETSO itself were eventually combined with the creation of the European Network of Transmission System Operators for Electricity (ENTSO-E) in 2009. The creation of ENTSO-E has externally empowered TSOs to overcome organisational fragmentation with the creation of a single EU-wide association. At the time of writing, the organisation comprises 43 electricity TSOs from 36 countries across Europe.

TSOs and energy governance in the EU

TSOs have been central actors in the process of developing a governance framework for the internal electricity market. Analysing the development of energy governance since the first electricity directive, we find that TSOs proactively shaped the emerging architecture, as summarised in Figure 1.

European regulation

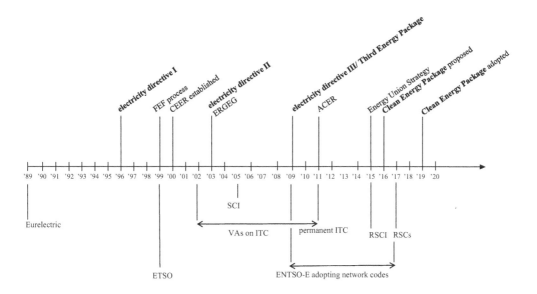

Figure 1. Timeline of European energy regulation and industry cooperation.
Sandra, E. 2009. Reproduced and adapted with permission of Palgrave Macmillan. Author's Illustration.

By the late 1990s, the European Commission initiated dedicated energy forums in order to build up policy expertise and competence. The Florence Electricity Forum (FEF), created in 1998, brought together market participants and regulators (Eberlein 2003; Vasconcelos 2001; Héritier 2003; Eberlein 2008). TSO as well as regulatory cooperation originated in the Florence process, triggering the creation of ETSO in 1999, and the Council of European Regulators (CEER) in 2000. The second electricity directive attributed a formal role to the regulators meeting in the European Regulators Group for Electricity and Gas (ERGEG, created by Commission decision 2003/796 EC). ERGEG was dissolved with the creation of the Agency for the Cooperation of Energy Regulators (ACER) in 2009. The third energy package also revamped TSO cooperation by giving them a formal mandate meeting as the European Network of Transmission System Operators for Electricity (ENTSO-E).

It is fair to say that thanks to their legacy of successful cross-border cooperation TSOs were initially trusted more by the Commission than the regulators, the latter being newly created institutions in most of the member states (Interview ETSO 2007). This still seemed to be the case at the time the third energy package was discussed. In its initial proposal the European Commission had envisaged a strikingly weak role for the regulators' network, to function mainly as a consulting body with hardly any direct competencies, while giving ENTSO-E a strong role (European Commission 2007, 11–2, 25–7). The compromise ultimately struck enhanced regulatory oversight with the creation of ACER. However, still in 2016 the Commission emphasised that the agency's function '[...] is not the execution of delegated regulatory Commission competencies, but the coordination of the regulatory decisions of independent national regulators' (European Commission 2016a, 22). A fourth package of energy legislation was proposed by the European Commission in November 2016 under the heading 'Clean Energy for all Europeans', the adoption of which was completed in May 2019. The major issue for TSOs was the proposal to introduce so-called Regional Operational Centers (ROCs) in order to enhance regulatory oversight at regional level. TSOs were fiercely opposed to the idea of introducing another layer of regulation and instead posited that the existing Regional Security Coordinators (RSCs) were sufficient to address security of supply issues, and argued for a forum process (ENTSO-E 2017). In the end, policymakers in 2019 agreed to introduce Regional Coordination Centres (RCCs) replacing the Regional Security Coordinators (RSCs). TSOs had first created the Security Coordination Initiative (SCI) in 2005, which was transformed into the Regional Security Coordination Initiatives (RSCIs) in 2015, and institutionalised as Regional Security Coordinators (RSCs) in 2017. European legislation thus formalises pre-existing formats of TSO cooperation.

The dynamics of authority displacement on cross-border issues

Having made a general argument about the level of authority acquired by TSOs, we now turn to present our case study analysis. We seek to discern the dynamics of authority displacement in further detail. We find that three settings of authority displacement towards TSOs materialise in the internal energy market (summarised in Table 2): (1) the *Escape Route* in the case of developing the network codes; (2) the *Political Route* involving soft coordination in the area of network planning; (3) the *Technical Route* involving self-regulation of TSOs on inter-TSO compensation (ITC). By contrast, we do not find a setting which corresponds to (4) the *Technocratic Route*. Solely relying on administrative authority is not an option for infrastructure regulation, due to a lack of technical and operational expertise.

Closing regulatory gaps via delegated private authority: the escape route

The third energy package gave TSOs important powers in developing network codes that are legally binding. The Commission, in its initial proposal, had suggested giving TSOs an even more encompassing mandate. The development of EU-wide market and technical codes formed part of this mandate (European Commission 2007, 13–5), as DG Energy and Transport was of the opinion that only TSOs held the required expertise to develop these

Table 2. Authority displacement towards TSOs in the internal electricity market.

Authority Shift	Strong	Weak
Sovereignty-based Contestation		
Strong	**Network Codes** (1) *Escape Route*: closing regulatory gaps via delegated private authority	**Network Planning** (2) *Political Route*: political control upheld through soft coordination
Weak	**Inter-TSO Compensation** (3) *Technical Route*: status quo orientation endorses entrepreneurial private authority	**[No example]** (4) *Technocratic Route*: administrative authority lacks technical and operational expertise

Source: own table.

codes. This view was subsequently contested by MEPs who argued that the agency, rather than the TSOs, should be in charge of developing technical codes in areas such as trading and market transparency and consulting stakeholders (European Parliament 2008, 33–4). However, the move to a stronger agency was opposed by member states who contested further shifts of regulatory authority towards the European level. Empowering TSOs thus appeased both the European Commission and the Council, while MEPs still managed to secure more regulatory oversight granted to the new agency. A lateral shift towards private authority thus helped to avoid gridlock, while also building on a legacy of entrepreneurial authority. Creating ENTSO-E and giving it a mandate to draft network codes constituted an *Escape Route* for European policymakers.

The process of formulating network codes (defined in articles 6 and 8 of regulation (EC) 714/2009) involves a strong element of delegated private authority. In fact, ENTSO-E has become a "fundamental regulatory agent" (Mathieu 2016, 113). First, on request by the Commission, and following consultation of stakeholders, ACER formulates a framework guideline. Second, TSOs conduct stakeholder consultation and draft network codes. Third, ENTSO-E sends the draft network code to ACER. Fourth, ACER writes an opinion and sends the codes to the Commission. Fifth, the Commission sends the codes to an Electricity Cross-Border Committee of specialists from national energy ministries who adopt the codes, with approval of the Council and the EP. Finally, the codes become binding, directly applicable regulation.

Contestation of the TSOs' power in drafting the network codes is mainly substance-based on the grounds that network operators holding a share of regulatory power use the procedure to their own economic benefit. Based on our interviews such contestation mainly relies on conjecture rather than hard facts. It seems to be too costly or even impossible to obtain a neutral assessment of the current situation (Interview COM 2017, 2019a). During the negotiations of the Clean Energy Package demands were voiced to introduce changes to network code development in order to allow for more transparency and checks and balances (Eyre 2016, 144). The electricity regulation (regulation (EU) 2019/943) as part of the Clean Energy Package has introduced some changes to the procedure which seek to enhance regulatory oversight. ACER can propose amendments to the codes, and has the right to revise the draft network codes submitted by ENTSO-E. To what extent these changes will *de facto* provide for more regulatory control has yet to be seen.

Upholding political control through soft coordination: the political route

Authority in the area of network planning remains very much in the national remit. This results from considerable sovereignty-based contestation of any attempts to shift network-planning competencies towards the European level. National governments prefer to safe-guard political control over an issue that has immediate consequences – costs as well as benefits – for their national constituencies. Compared to other, more technical issues of infrastructure governance, the construction of new infrastructure is a highly visible project, often controversial and involving legal disputes regarding processes of authorisation.

As a consequence, European governance is mainly about soft coordination to achieve goals set at the aggregate EU level. The EU is striving to achieve a ten percent inter-connection target by 2020, which is expected to increase to 15% by 2030. The Commission first identified projects in its 2006 priority interconnection plan (European

Commission 2016b). Since 2013, a list of so-called Projects of Common Interest (PCIs) has been compiled on a biannual basis, following impact assessment and comprehensive consultation. The realisation of PCIs is supported through faster permit procedures and the right to apply for EU funding lines, provided under the Connecting Europe Facility between 2014 and 2020. The Commission has further created an Energy Infrastructure Forum that meets annually in the context of the Energy Union strategy. The Forum brings together a wide range of stakeholders and covers issues related to network development such as cross-border cost allocation, innovation and public acceptance (European Commission 2017). While such coordination can lead to visibly set targets across the EU, it is ultimately national planning, subject to approval by national regulators, that matters. In this issue area decision-making, therefore, follows the *Political Route*, involving a strong degree of political contestation and a weak shift towards either supranational or private authority.

Private actors generate evidence and prepare substantive policy input, but they can only formulate desiderata. Every 2 years ENTSO-E proposes a Ten Year Network Development Plan (TYNDP). TSOs submit a draft plan to the Agency for its opinion concerning its consistency with national planning. ACER can recommend amendments to the Commission on both types of plans. Usually, ENTSO-E has accepted too many proposals, which follows from the TSOs' economic rationale to invest in their asset base and interconnectors (Eyre 2016, 137). TSOs can indeed pass on the cost of constructing new infrastructure, while being able to defend their rationale by putting forward security of supply arguments (Interview COM 2017, 2019b). TSO claims are, however, subject to substance-based contestation both at the national and European level. National regulatory authorities authorising the projects in the member states will seek to reduce the cost, and do not grant permission easily. At EU-level ACER has started to voice concerns about the ENTSO-E perspective, arguing that TSOs exaggerate the need for new capacity. The agency further argues that TSOs could make more efficient use of existing capacity, for instance through up-to-date software tools (Interview COM 2017, 2019a). In the Clean Energy Package the Commission urges TSOs to '[…] conduct an extensive consultation process, and at an early stage and in an open and transparent manner, involving all relevant stakeholders' (European Commission 2016a, 62). The TSO role of providing evidence in the area of network planning therefore not only meets national resistance (sovereignty-based contestation), but also substance-based contestation.

Entrepreneurial authority on ITC: the technical route

Finally, TSOs hold entrepreneurial authority in the area of inter-TSO compensation (ITC). Here we have a setting where sovereignty-based contestation has faded away over time, since the political benefits of exerting control proved limited. It became possible to entrust TSOs with a task framed as being mainly of technical relevance (*Technical Route*).

Remuneration schemes were historically designed within the remits of nation states, yet electricity flows do not stop at borders. An appropriate mechanism to cover the cost of transmission losses and the infrastructure cost of hosting transit flows thus had to be found. This was one of the priorities for the European Commission when launching the Internal Energy Market. Lacking the required technical expertise, information and regulatory capacity, the Commission asked the Florence Energy Forum created in 1998 to come up with a solution. In the subsequent discussions the issue turned out to be highly controversial, and decisionmaking was stalled several times (Eberlein 2003). While policymakers and regulators had difficulties agreeing on a common framework, infrastructure operators began to take measures unilaterally. To use Jessica Green's conceptualisation, TSOs thereby acquired 'entrepreneurial authority' in confronting governance failure (Green 2014, 17). Between 2002 and 2010, European TSOs negotiated several agreements within their European association ETSO. These agreements involved a process of continuous renegotiation (ITC schemes for 2002, 2003, 2004, 2005–2007, 2007, 2008–2009, 2010). In the second energy package, the Commission reiterated the need for guidelines to govern voluntary coopera- tion between TSOs. The process to agree on such guidelines proved difficult, given that the NRAs inside ERGEG disagreed significantly on the appropriate design of ITC schemes. In the context of the Third Energy Package TSOs were required to come up with a permanent scheme. Since May 2011 a permanent ITC mechanism has been put into place. This formalisation of pre-existing voluntary cooperation involves a degree of regulatory over- sight: ACER submits a proposal for a framework fund to compensate TSOs for the cost of additional infrastructure required to host transit flows, on which the Commission decides. The agency also formulates a proposal on how to compensate for the provision of infra- structure for cross-border flows. Despite such adjustments, the ITC scheme, in essence, reproduces the same model previously used by TSOs for their voluntary agreements. Entrepreneurial authority hence has received endorsement by policymakers.

That said, there is some substance-related contestation of the ITC scheme, voiced by participating TSOs and regulators. Through inter-TSO compensation, TSOs have to solve a redistributive conflict, which is why it does not come as a surprise that some participat- ing TSOs are structurally at a disadvantage, namely those that pay more as a result of one method being chosen over another. Regulators criticise the scheme, claiming that it does not provide enough incentives to develop the network efficiently. This was the line of argumentation put forward in a March 2013 ACER recommendation on ITC (ACER 2013). Despite such criticism, there seems to be an overall preference to maintain the status quo (Interview regulator 2017; Interview COM 2017, 2019a).

Conclusions

In our case study, we discern three distinct ways in which authority has been displaced. First, in the case of network codes, we look at an intriguing setting where despite a high level of contestation we see a strong shift towards private authority. We argue that this outcome can be explained as a strategy to escape deadlock. Sovereignty-based contesta- tion has made a substantially stronger European agency impossible, yet a widening regulatory gap needed to be filled. In order to avoid the looming governance failure, policymakers embarked on an *Escape Route* which allowed TSOs to acquire a significant share of the regulatory authority in the process of developing network codes. There is some substantive criticism of this regulatory role, based on suspicions that infrastructure operators generate undue economic benefits as a result. Yet, there is a lack of substantiat- ing evidence for such claims. The Clean Energy Package fine-tunes the existing architec- ture and enhances regulatory oversight, but maintains the role for TSOs.

Second, in the case of network planning sovereignty-based contestation of a shift towards the European level has been strong. Member states see political benefits in keeping the upper hand in infrastructure decisions. At the same time, we observe a weak shift towards private authority at the European level, notably due to acquired levels of regulatory oversight in the national remit. We argue that the mechanisms of soft coordination, which leave national sovereignty largely untouched, constitute a likely outcome in a politicised setting. This *Political Route* is therefore very much in line with the desire to safeguard national sovereignty.

Third, in the case of inter-TSO compensation, sovereignty-based contestation has faded away over time, while we observe a strong simultaneous shift towards private authority. This is the *Technical Route* of integration, which endorses entrepreneurial authority in functionally driven cross-border cooperation. Here we also detect a degree of substance-based contestation, which is, however, overshadowed by an overall preference to maintain the status quo.

Finally, we do not find evidence for a *Technocratic Route*, which would involve a low level of sovereignty-based contestation allowing for delegation to administrative authority. Here we can only speak to our case-study, which is about infrastructure governance. In this particular area of regulation, public actors rely, to a significant extent, on the technical expertise and the operational capacity of private actors. Beyond the scope of our case study other articles in this Special Issue provide further evidence for the contested and politicised nature of energy policy, which make the technocratic route even less of an option.

With its focus on private authority, our analysis makes a valuable contribution to existing research on European integration in the field of energy. The lateral shift of authority towards private actors has so far been largely overlooked in the literature. We find that in the new context of integrated markets, the TSOs' long-standing role of providing cross-border capacity for the flow of energy gives these organisations new power positions as guarantors of both trade flows and of security of supply. Drawing on their legitimacy as 'connectors' in the EU electricity market, network operators have been able to shape the emerging European policy framework and acquire a regulatory role formalised by European legislation. While empowering private actors, rather than supranational agents, has helped to circumvent political resistance to completing the internal energy market, it reinforces the lack of effective regulatory oversight at the transnational and European levels. We conclude that despite substantial lateral shifts of authority towards TSOs the level of contestation is strikingly low. This shows that hidden integration, framed as a matter of technical and operational cooperation, can be effectively sheltered from public attention.

Acknowledgments

The authors are grateful to the academic association for Contemporary European Studies (UACES) for the funding of the *Collaborative Research Network on European Energy Policy* (2015–2018), which was the breeding ground for this special issue. We are also grateful for *Universiteitsfonds Limburg* (SWOL) for co-funding the authors' workshop in Maastricht, in April 2018. Research for this article was supported by the European Union's Sixth Framework Programme (FP6-CITIZENS, New Modes of Governance NEWGOV, grant agreement ID 506392, Project 5 "New Modes of Governance in the Shadow of Hierarchy") and by the LOEWE Center on Sustainable Financial Architecture for Europe (Project "The State of the Union: The Politics of Integration in Banking and Energy", SAFE funding agreement #21136). Moreover, the article was researched and written during a sabbatical fellowship awarded by the Johanna Quandt Young Academy (JQYA) to Sandra Eckert in the period from October 2018 to February 2019. The article was finalised when Sandra Eckert held a COFUND-AIAS fellowship (European Union Horizon 2020 Research and Innovation Programme Marie Sklodowska-

Curie grant agreement no. 754513 and Aarhus University Research Foundation) awarded by the Aarhus Institute of Advanced Studies (AIAS) as of October 2019. We also thank several colleagues who provided detailed comments on our paper, with special thanks to Francesca Batzella, Katja Biedenkopf, Helene Dyrhauge and Rosa Fernández, as well as all the colleagues of the UACES CRN on European Energy Policy. Moreover, we are indebted to the editors of this special issue, an anonymous reviewer and the JEI editors for their helpful comments and guidance. Finally, we thank Amber Davis for providing excellent proofreading services.

Disclosure statement

No potential conflict of interest was reported by the authors.

Funding

This work was supported by the Aarhus Universitets Forskningsfond;European Union FP6-CITIZENS, New Modes of Governance NEWGOV [ID 506392];LOEWE Center on Sustainable Financial Architecture for Europe [#21136];Johanna Quandt Young Academy;European Union Horizon 2020 Research and Innovation Programme Marie Sklodowska-Curie [754513].

References

ACER. 2013. *A New Regulatory Framework for the Inter-Transmission System Operator Compensation*. Recommendation of the Agency for the Cooperation of Energy Regulators 05/2013. 25 March. Ljubiljana: Agency for the Cooperation of Energy Regulators.

Bjørnebye, H. 2006. "Interconnecting the Internal Electricity Market: A Goal without A Plan?" *Competition and Regulation in Network Industries* 1 (3): 333–353. doi:10.1177/178359170600100301.

Bocquillon, P., and T. Maltby. forthcoming. "EU Energy Policy Integration as Embedded Intergovernmentalism: The Case of Energy Union Governance Regulation." *Journal of European Integration* 42 (1), forthcoming.

Boswell, C. 2009. *The Political Uses of Expert Knowledge: Immigration Policy and Social Research*. Cambridge: Cambridge University Press.

Commission, European. 2017. "Communication on Strengthening Europe's Energy Networks." In *Com (2017) 718 Final*. Brussels: European Commission.

Consolini, P. M., and T. R. LaPorte. 1991. "Working in Practice but Not in Theory: Theoretical Challenges of 'High-reliability Organizations'." *Journal of Public Administration Research and Theory* 1 (1): 19–48. doi:10.1093/oxfordjournals.jpart.a037070.

Crisan, A., and M. Kuhn. 2016. "The Energy Network: Infrastructure as the Hardware of the European Union." In *Energy Union: Europe's New Liberal Mercantilism?* edited by S. S. Andersen, A. Goldthau, and N. Sitter, 165–182. London: Palgrave Macmillan.

Cutler, C. A., V. Haufler, and T. Porter. 1999. *Private Authority and International Affairs*. Albany, NY: Suny Press.

Eberlein, B. 2003. "Regulating Cross-Border Trade by Soft Law? the "Florence Process" in the Supranational Governance of the Electricity Markets." *Journal of Network Industries* 4 (2): 137–155.

Eberlein, B. 2008. "The Making of the European Energy Market: The Interplay of Governance and Government." *Journal of Public Policy* 28 (1): 73–92. doi:10.1017/S0143814X08000780.

Eberlein, B. 2012. "Inching Towards a Common Energy Policy: Entrepreneurship, Incrementalism, and Windows of Opportunity." In *Constructing a Policy-Making State? Policy Dynamics in the EU*, edited by J. Richardson, 147–169. Oxford: Oxford University Press.

Eberlein, B., and E. Grande. 2005. "Beyond Delegation: Transnational Regulatory Regimes and the EU Regulatory State." *Journal of European Public Policy* 12 (1): 89–112. doi:10.1080/1350176042000311925.

Eckert, S. 2011. "European Regulatory Governance." In *Handbook on the Politics of Regulation*, edited by D. Levi-Faur, 513–524. Cheltenham: Edward Elgar Publishing.

Eckert, S. 2016. "The Governance of Markets, Sustainability and Supply. Toward a European Energy Policy." *Journal of Contemporary European Research* 12 (1): 502–517.

Eckert, S. 2019. *Corporate Power and Regulation. Consumers and the Environment in the European Union*. London: Palgrave.

ENTSO-E. 2017. "Power Regions for the Energy Union: Regional Energy Forums as the Way Ahead." *Brussels* 13 (October): 2017.

ENTSO-E. 2018. "Frequency Deviations in Continental Europe - Discussions Ongoing to Find a Solution." *Brussels* 12 (November): 2018.

EurActiv. 2018. "European Clocks Slowed by Serbia-Kosovo Power Grid Row." Accessed 08 March 2018. https://www.euractiv.com/section/enlargement/news/european-clocks-slowed-by-serbia-kosovo-power-grid-row/

European Commission. 2007. *Proposal for a Regulation of the European Parliament and of the Council Establishing a European Union Agency for the Cooperation of Energy Regulators*. COM (2007) 530 final. Brussels: European Commission.

European Commission. 2016a. *Proposal for a Regulation of the European Parliament and of the Council on the Internal Market for Electricity*. COM (2016) 861 final. Brussels: European Commission.

European Commission. 2016b. *Communication of 10 January 2007 from the Commission to the Council and the European Parliament Entitled "Priority Interconnection Plan"*. COM (2006) 846 final. Brussels: European Commission.

European Commission. 2017. *Communication on strengthening Europe's energy networks*. COM (2017) 718 final. Brussels: European Commission.

European Parliament. 2008. *Report on the Proposal for a Regulation of the European Parliament and of the Council Amending Regulation (EC) No 1228/2003 on Conditions for Access to the Network for Cross-Border Exchanges in Electricity (COM (2007)0531-C6-0320/2007-2007/0198(COD)) A6-0228/ 2008*. Brussels: Committee on Industry, Research and Energy.

Eyre, S. 2016. "An Industry Perspective: The Primacy of Market-Building." In *Energy Union: Europe's New Liberal Mercantilism?* edited by S. S. Andersen, A. Goldthau, and N. Sitter, 133–146. London: Palgrave Macmillan.

Fischer, F. 1990. *Technocracy and the Politics of Expertise*. Newbury Park, CA: SAGE Publications.

Fuchs, D., T. Gumbert, and B. Schlipphak. 2017. "Eurosecpticism and Big Business." In *The Routledge Handbook of Euroscepticism*, edited by B. Leruth, N. Startin, and S. Usherwood, 317–330. New York: Routledge.

Glachant, M. 2003. "Voluntary Agreements Under Endogenous Legislative Threats." *FEEM Working Paper* No. 36, Milan.2003.

Green, J. F. 2014. *Rethinking Private Authority. Agents and Entrepreneurs in Global Environmental Governance*. Princeton: Princeton University Press.

Halfteck, G. 2006. *A Theory of Legislative Threats*. Tel Aviv: Tel Aviv University.

Halfteck, G. 2008. "Legislative Threats." *Stanford Law Review* 61: 629–710.

Héritier, A. 1999. *Policy-making and Diversity in Europe: Escaping Deadlock*. Cambridge: Cambridge University Press.

Héritier, A. 2003. "New Modes of Governance in Europe: Increasing Political Capacity and Policy Effectiveness?" In *The State of the European Union. Law, Politics, and Society*, edited by T. A. Börzel and R. A. Cichowski, 105–126. Oxford: Oxford University Press.

Héritier, A., and S. Eckert. 2008. "New Modes of Governance in the Shadow of Hierarchy: Self-Regulation by Industry in Europe." *Journal of Public Policy* 28 (1): 113–138. doi:10.1017/S0143814X08000809.

Herranz-Surrallés, A., I. Solorio, and J. Fairbrass. forthcoming. "Renegotiation Authority in the Energy Union: A Framework for Analysis." *Journal of European Integration* 42 (1), forthcoming.

Hood, C. 2002. "The Risk Game and the Blame Game." *Government and Opposition* 37 (1): 15–37. doi:10.1111/1477-7053.00085.

Interview COM. 2017, 2019a. Grid expert DG Energy, Brussels, 12.04.2017 and 10.01.2019.

Interview COM. 2017, 2019b. National expert, DG Energy, Brussels, 11.04.2017 and 10.01.2019.

Interview ETSO. 2007. Former secretary general, Brussels, 17.12.2007.

Interviewregulator. 2017. Grid expert, OFGEM, former member ETNSO, London, 13.07.2017.

Jevnaker, T. 2015. "Pushing Administrative EU Integration: The Path Towards European Network Codes for Electricity." *Journal of European Public Policy* 22 (7): 927–947. doi:10.1080/13501763.2014.1000363.

Kahler, M., and D. A. Lake. 2004. "Governance in a Global Economy: Political Authority in Transition." *Political Science and Politics* 37 (3): 409–414. doi:10.1017/S1049096504004573.

Knill, C. 2001. "Private Governance across Multiple Arenas: European Interest Associations as Interface Actors." *Journal of European Public Policy* 8 (2): 227–246. doi:10.1080/13501760110041569.

Labelle, M. C. 2016. "Regulating for Consumers? the Agency for Cooperation of Energy Regulators." In *Energy Union: Europe's New Liberal Mercantilism?* edited by S. S. Andersen, A. Goldthau, and N. Sitter, 147–164. London: Palgrave Macmillan.

Lodge, M. 2008. "Regulation, the Regulatory State and European Politics." *West European Politics* 31 (1): 280–301. doi:10.1080/01402380701835074.

Majone, G. 1996. *Regulating Europe*. London, NY: Routledge.

Mathieu, E. 2016. *Regulatory Delegation in The European Union: Networks, Committees and Agencies*. London: Palgrave Macmillan.

Mathieu, E., and B. Rangoni. 2019. "Balancing Experimentalist and Hierarchical Governance in European Union Electricity and Telecommunications Regulation: A Matter of Degrees." *Regulation & Governance* 13 (4): 577–592. doi: 10.1111/rego.v13.4.

Meletiou, A., C. Cambini, and M. Masera. 2018. "Regulatory and Ownership Determinants of Unbundling Regime Choice for European Electricity Transmission Utilities." *Utilities Policy* 50: 13–25. doi:10.1016/j.jup.2018.01.006.

Pfeffer, J., and G. R. Salancik. 2003. *The External Control of Organizations: A Resource Dependence Perspective*. Stanford: Stanford Business Books.

Quack, S. 2016. "Expertise and Authority in Transnational Governance." In *Authority in Transnational Legal Theory. Theorising across Disciplines*, edited by R. Cotterrell and M. Del Mar, 361–387. Cheltenham: Edward Elgar.

Rangoni, B. 2019. "Architecture and Policy-making: Comparing Experimentalist and Hierarchical Governance in EU Energy Regulation." *Journal of European Public Policy* 26 (1): 63–82. doi:10.1080/13501763.2017.1385644.

Rottmann, K., and A. Lenschow. 2008. "'privatising' EU Governance: Emergence and Performance of Voluntary Agreements in European Environmental Policy." In *Multi-Level Governance in the European Union: Taking Stock and Looking Ahead*, edited by T. Conzelmann and R. Smith, 232–254. Baden-Baden: Nomos.

Sandra, E. 2019. *Corporate Power and Regulation. Protecting Consumers and the Environment in the European Union*. London: Palgrave Macmillan.

Sending, O. J. 2015. *The Politics of Expertise: Competing for Authority in Global Governance*. Michigan: University of Michigan Press.

Thatcher, M., and A. S. Sweet. 2002. "Theory and Practice of Delegation to Non-Majoritarian Institutions." *West European Politics* 25 (1): 1–22. doi:10.1080/713601583.

Tosun, J., and M. Mišić. forthcoming. "Conferring Authority in the European Union: Citizens' Policy Priorities for the European Energy Union." *Journal of European Integration* 42 (1), forthcoming.

Vasconcelos, J. 2001. "Cooperation between Energy Regulators in the European Union." In *Regulation of Network Utilities. The European Experience*, edited by C. Henry, M. Matheu, and A. Jeunemaître, 284–289. Oxford: Oxford University Press.

Contested energy transition? Europeanization and authority turns in EU renewable energy policy

Israel Solorio and Helge Jörgens ⓘ

ABSTRACT

In a context of multiple crises, the European Union's climate and energy policies have become highly politicized and contested. Based on a comparative study of renewable energy policies in ten EU member states, and adopting a circular view of policy change and Europeanization to account for overlapping sovereignty claims between the national and the European level, this article unravels the authority debates over successive rounds of negotiation, adoption, and implementation along three EU directives. Following an exploratory process-tracing method, we investigate how policy-making authority originally delegated to the EU becomes contested by the member states and how these authority conflicts are managed. We find that the Europeanization of renewable energy policy is accompanied by an issue-specific renegotiation of authority between the EU and its member states which, in times of crises, can trigger instances of de-Europeanization and even a partial weakening of European integration in this policy domain.

Introduction

For decades, ever since the publication of the European Commission's White Paper on renewable energy sources (RES) in 1997, RES promotion has been acknowledged as a landmark component of EU climate and energy policies. Its relevance can be explained by a mix of factors related to the EU's institutional structure and the resulting nature of these policies. On the one hand, before a formal competence on energy was granted to the EU under the Treaty of Lisbon, environmental policymaking – together with internal market policies – provided a means for increasing EU participation in this policy domain (Tosun and Solorio 2011). As a result, RES promotion became one of the most effective ways to shape national energy policies. On the other hand, the EU's aspiration to become an international leader in climate change has placed further pressure on European policymaking to pursue an ambitious internal climate policy, with RES promotion being one of its pillars.

For many years, the EU's authority in this sub-field of climate and energy policies remained largely uncontested due to a consensus among decision-makers at all levels about the environmental, economic, security and social advantages of RES. However, in the context of multiple crises, tensions in EU renewable energy policy have emerged and the socio-economic benefits of the energy transition are increasingly contested. If the 2009 Renewable Energy Directive (RED) containing the goals towards 2020 was passed with a broad support from member states, the road to the 2030 goals was more difficult and crowded with authority claims on the part of national governments. Against this background, the 2030 climate and energy framework agreed at the European Council of October 2014, with its greenhouse emission reduction emphasis, has been interpreted as an indicator that RES promotion is being sidelined from EU priorities (Bürgin 2015; Solorio and Bocquillon 2017). Whilst the adoption of the 2018 directive – known as RED II – brought the EU's energy transition back on track, its difficult negotiations and lack of consensus among member states revealed that conflicts of authority are more than present in this policy.

This article disentangles the renegotiation of authority in EU renewable energy policy. Focusing on the emergence and change of renewable electricity (RES-E[1]) policies in the EU and its member states, it explores: (i) how and why authority was conferred on the EU; (ii) what types of contestation on the part of member states have emerged; and (iii) the ways in which authority conflicts have been addressed. To answer these questions this article is guided by the debates on authority contestation in the European multilevel polity as well as the literature on circular Europeanization of public policies. It develops a longitudinal analysis which traces the negotiation, adoption, and implementation of each of the three key legislative pieces for RES promotion: the 2001 RES-E directive, the 2009 RED, as well as the most recent 2018 RED II. The developed analysis focuses on two main features of this policy and its evolution over time: the nature of targets, which impacts on the EU's capacity to monitor compliance, and the debate about the support schemes, which relates to the EU's authority to determine the means of RES promotion in the member states.

The article is structured as follows. Section 2 presents our analytical framework, which brings together this Special Issue's focus on the *renegotiation of authority* with recent debates on the *Europeanization* of member state policies. The methodology is presented in Section 3. Section 4 introduces the authority debates in EU renewable energy policy that emerged during the period of observation (2001–2018). In each stage the way authority is conferred, the sources of authority contestation, the management of conflicts and the effects of policy implementation at the national level are analyzed. The findings are discussed in Section 5, where conclusions are presented.

Analytical framework: feedback loops and the renegotiation of authority in the EU

One cannot study the development and change of public policies in EU member states without taking into account the specific nature of the EU multilevel system. The vast Europeanization literature that has evolved over the past decades does exactly this by asking 'if and how the EU has changed representation, governance and public policy in the member states and beyond.' (Radaelli and Exadaktylos 2010, 189). Having become increasingly sophisticated and rigorous, Europeanization presents itself as a useful diagnostic framework for exploring the vertical displacements of authority between the EU and its member states that are at the core of this Special Issue. Today, the Europeanization literature provides useful models and analytical tools for furthering our understanding of how 'EU institutions and policies are becoming more politicized and contested domestically' (see Herranz-Surrallés, Solorio, and Fairbrass forthcoming) and how authority is being renegotiated between the EU and its member states. Its explanatory potential has been particularly enhanced with the inclusion of a circular perspective (Saurugger 2014), which considers 'feedback loops' in order to facilitate the observation not only of the reasons behind the delegation of authority upwards to the EU (bottom-up Europeanization) and the changes derived from the EU's impact at the national level across time (top-down Europeanization), but also of the salience and politicization of issues as well as the extent to which member states' governments adopt supportive or critical positions towards European integration (Saurugger and Radaelli 2008).

Early understandings defined Europeanization as a 'two-way process', including the uploading and downloading of policies (Börzel 2002, 193), where member states first try to actively shape the form and content of European Integration in order to subsequently 'maximize the benefits and minimize the costs' of adapting to it (Börzel 2002, 196). Gradually, this approach evolved and a stronger emphasis was placed on the role of domestic actors, both for influencing national positions in the negotiations at the EU level and during the national adaptation to EU pressures. Regarding policy implementation, this shift resulted in the notion of Europeanization as 'usage', arguing that it is crucial to investigate the ways in which domestic actors seize opportunities and work around Europeanization constraints (Woll and Jacquot 2010). Subsequent contributions included a horizontal dimension of Europeanization, where member states directly influenced each others' policies within the institutional structure of the EU (Bulmer and Padgett 2005). This development came together with the recognition that European policymaking is not necessarily based on the EU-wide standardization of regulations, but in some policy domains includes a strong reliance on softer modes of governance (Treib, Holger, and Falkner 2007), which can be interpreted as a way to prevent or manage authority conflicts between member states and EU institutions (see Herranz-Surrallés, Solorio, and Fairbrass forthcoming).

Drawing on these insights, and motivated by the centrifugal effects of crisis on European integration, more recent approaches capture the European policy-process more interactively. In this way, Europeanization not only produces changes domestically but can actually generate political disagreement, which in turn can lead domestic actors to mobilize either for or against subsequent instances of supranational governance (Coman 2014). National preferences, thus, are not static but determined by the new equilibria generated by previous rounds of Europeanization. This can lead to the paradoxical situation where Europeanization can be the cause of de-Europeanization, understood as a practice in which a member state 'de-constructs previous advancements made through the process of Europeanization' (Copeland 2016, 1126).

Figure 1 synthesizes this circular model of EU policymaking, which noticeably is strongly interlinked to authority debates in the EU. Its basic assumption is that EU policymaking occurs in cycles where the domestic 'downloading' of EU policies is not the end point, but potentially also the start of a new round of circular Europeanization. Starting with the delegation of authority, bottom-up Europeanization sheds light on the processes of conferring authority vertically on the EU. Following the policy cycle, top-down Europeanization is at the same time an explanatory variable for domestic policy change and a factor that can either stifle or provoke the contestation of authority by

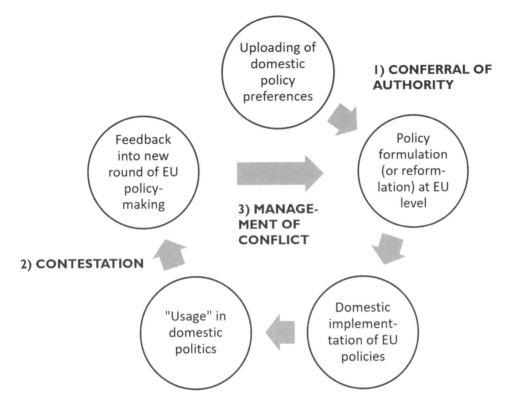

Figure 1. A model of circular Europeanization.
Source: Authors' illustration.

member states. Managing authority conflicts is a key requisite to close the circle of the policy process and to start a new round of Europeanization.

In line with the framework of the Special Issue (see Herranz-Surrallés, Solorio, and Fairbrass forthcoming), we assume that a conferral of authority may occur because of functional needs or be driven by value-based objectives. In addition, claims over authority can be categorized as sovereignty-based and substance-based contestation. Finally, the management of authority conflicts can pursue either legal or political strategies.

We expect EU renewable energy policy to be a suitable case for assessing policy change over time, examining how Europeanization generates winners and losers at the domestic level (e.g. new vs. old RES companies) and for analyzing how this, in turn, changes national positions towards subsequent rounds of European Integration. Regarding the time span, the period between 2001–2018 is sufficient for observing feedback loops involving higher levels of politicization and contestation of EU authority after the 2008–09 financial and economic crises, and the resulting new European governance arrangements.

Methods

The determination of causality has been a constant challenge for Europeanization research. One problem is that most Europeanization studies rely on single case designs and lack a clear 'justification of case selection' (Exadaktylos and Radaelli 2012, 31). More importantly, despite being a major research field, Europeanization studies have been weak in generating clearly defined and testable hypotheses about why, when and under which conditions developments at the European level lead to policy change at the national one. Instead of developing theories that are specific to the process of Europeanization, researchers usually 'draw on existing theories and models of comparative politics and international relations' (Radaelli 2018, 56). When applying these theories to the Europeanization of national policies, the resulting causal hypotheses necessarily fail to account for the multiplicity of potentially relevant explanatory and intervening variables. Complex research designs based on the notion of circular Europeanization, where both European policies and domestic adaptation can be either independent or dependent variable, further heighten the challenge (Saurugger 2014). Here, domestic actors play the role of intervening variables, both for the national adaptation to EU pressures and for influencing national positions during the negotiations at the EU level.

Considering methodological challenges such as complex causality as well as the need to understand the role played by procedural factors including recurrent instances of policy formulation and implementation and feedback loops between different rounds of policymaking (Rohlfing 2012), we have chosen an exploratory rather than a hypothesis-testing approach. Exploratory process tracing is an adequate method in areas where precise, testable and theory-based hypotheses are scarce or do not exist. It allows for a systematic analysis of policy processes over time, taking into account a number of factors that are of potential importance for the observed outcome. The aim is to develop expectations about potential causal mechanisms which then can serve as a theoretical basis for future research, an approach which Rohlfing (2012, 41) refers to as an 'exploratory case study centered on outcome'.

The study presented here is based on ten qualitative country studies on the Europeanization of domestic RES policies (Bulgaria, Denmark, France, Germany, Italy, the Netherlands, Poland, Romania, Spain and the UK) and an analysis of RES policy developments at the EU level (Solorio and Jörgens 2017). The cases were selected on the basis of their expected roles as either leaders, laggards, or fence-sitters with regard to the promotion of RES-E (Liefferink and Andersen, 1998; Börzel 2002). Due to their roles as early adopters of RES-E policies, we expect Germany, the Netherlands, Denmark, and Spain to act as pace-setters in the Europeanization of RES-E policies. Based on their generally weak record of implementing EU environmental policies and their roles as late-comers with regard to RES promotion, we expect Italy, Poland, Romania and Bulgaria to take a more reluctant or oppositional stance. Finally, considering their ambivalent position towards RES promotion, we expect the UK and France to act as fence-sitters. This analysis comprises three consecutive legislative policy initiatives in this policy area – or Europeanization rounds – in order to account for feedback loops and the assumed circular character of Europeanization processes.

Authority turns in EU renewable energy policy

The development of EU renewable energy policy started with the 1973 oil crisis, which led European institutions to search for solutions to the Community's energy problems (Tosun and Solorio 2011). This initial phase consisted mostly of modest support for RES research and development and a soft coordination approach, with no significant relocation of authority from the national to the supranational level (Solorio and Bocquillon 2017, 24–25). Nevertheless, it is important to note that these early years determined the different national starting positions on RES promotion (Reiche and Bechberger 2004, 844) and, although marginal in terms of EU policy-making, this phase to some extent affected the subsequent negotiation positions of member states towards the RES-E directive.

The first round of circular Europeanization: the 2001 RES-E directive and the displacement of authority upwards

Conferring authority

The 1997 White Paper on RES promotion is commonly recognized as the moment where a distinct renewable energy policy domain began to emerge (Solorio and Bocquillon 2017, 25–26). It outlined a set of measures to overcome obstacles for RES development and established an indicative target of 12% RES in EU primary energy consumption by 2010. In 2000, the Commission followed up on the RES White Paper and proposed the first EU legislation explicitly oriented towards RES promotion, the RES-E directive. During the negotiation process, conflicts over the RES targets and the support schemes for RES promotion arose among EU institutions and the member states (Rowlands 2005). Regarding the targets, both the Commission and the European Parliament defended the need for mandatory national RES goals. In contrast, the Council considered that the indicative target of 12% was a sufficient guide for national efforts towards RES promotion (Meyer 2003). The dispute was settled with an agreement where targets, 'although relatively ambitious, remained non-binding' (Solorio and Bocquillon 2017, 27). Regarding support schemes, the Commission strongly pushed for harmonization by establishing a European market for trading in renewable energy certificates, a system known as Tradable Green Certificates (TGCs) (Busch and Jörgens 2012). This proposal was met with strong resistance from countries with Feed-in Tariffs (FITs) systems such as Germany and Spain, who advocated a general freedom on part of member states to choose their own support scheme.

The final RES-E directive had an indicative goal of 22.1% of RES-E in total EU electricity consumption by 2010 and included national indicative targets based on the technological and economic potential. The directive also included a provision that member states should publish periodically a report on their progress towards the national indicative targets. On that basis, the Commission had to assess the evolution towards the RES targets, both nationally and for the EU as a whole (Solorio and Bocquillon 2017, 26–27). Although there was no explicit mention of sanctions, the directive contemplated the possibility that, in case the Commission's assessment found national indicative targets to be inconsistent with the EU overall goal, it should present proposals to address this issue (including possible mandatory targets) (Rowlands 2005, 969–970). With this, a legal source of authority was given to the Commission to set RES goals in a context driven by the need to meet with the EU's Kyoto commitments. However, due to the lack of legal obligations and the absence of sanctions, the conferred authority was limited. Regarding support schemes, there was no prescription of a concrete EU model. Instead, the Commission was left in charge of reporting on the experience gained through the application and coexistence of different mechanisms (Busch and Jörgens 2012, 76). Member states maintained, thus, considerable leeway in the implementation of this policy. In addition, the directive also comprised a 'market-making' measure, oriented to remove grid connection barriers for RES-E. Overall, despite providing only a 'loose regulatory framework' (Solorio and Bocquillon 2017, 25), the RES-E directive represented a leap forward in the evolution of EU renewable energy policy.

Sources of contestation and the management of authority conflicts

Considering their experience in RES promotion, Denmark, Germany and Spain were natural candidates to lead the negotiation of the RES-E directive and, therefore, be able to upload their domestic policies – which were all, 'with some variations', based on the FITs model (Rowlands 2005, 971). Nevertheless, the Commissions' preference for more market-oriented support schemes based on TGCs made the UK, a late-comer in RES promotion, an 'unexpected pace-setter' given its traditional market approach to energy policy (Solorio and Fairbrass 2017, 108). In this context, the Commission tried to de-politicize the debate by using a 'competitive markets' argument. RES leaders adopted a defensive strategy, fighting against any attempt for an EU-wide harmonization of support schemes. Their success on RES promotion backed these positions, so the Commission had to give up its harmonization attempt (Rowlands 2005, 971–972). Where the RES leaders played their part was in setting an ambitious EU-wide RES goal. Denmark, Germany and Spain pushed for an ambitious EU target equivalent to 12% of the overall energy consumption; however, only Denmark and Germany wanted to make this target binding nationally (Rowlands 2005; Vogelpohl et al. 2017; Dyrhauge 2017). In sum, both for targets and for support schemes, soft governance was the solution to authority disputes during this stage of EU renewable energy policy.

Implementation and policy change

In terms of the impact of the RES-E directive at the domestic level, our cases show top-down Europeanization to be strongest before – and not after – the RES-E directive was adopted. For example, in Italy, the Bersani decree of 1999 introduced an ambitious support scheme for RES-E based on mandatory quotas and TGCs. A key driver for this was the publication in October 1998 of a first unofficial draft of the EU's 2001 RES-E directive, in which the European Commission expressed its strong preference for a support model based on quota systems (Di Nucci and Russolillo 2017, 127). It was this (rather remote) possibility of an EU-wide harmonization towards a system based on RES-E quotas and TGCs, and the resulting threat that support schemes based on FITs might cease to be in compliance with EU law, that triggered the Italian policy change. In France the negotiations of this directive built momentum for the inclusion of RES-E promotion in the Electricity Bill of 2000 (Bocquillon and Evrard 2017, 167–168). Something similar happened in Germany with the adoption in 2000 of the Renewable Energy Sources Act (Vogelpohl et al. 2017).

Moreover, during this phase, top-down Europeanization occurred in an indirect rather than a direct manner. In most countries, it was the EU-driven liberalization of the electricity markets rather than the RES-E directive that triggered the most important national policy changes. It did so by significantly changing domestic opportunity structures. On the economic side it removed structural and institutional barriers to market entry for producers of RES-E. On the political side, it set the course for domestic policies aimed at gradually increasing the share of RES-E in domestic energy production and consumption without raising the opposition of powerful 'natural' opponents such as the big power utilities or incumbents (Jörgens, Öller, and Solorio 2017, 290–292). During this stage, indirect top-down Europeanization through the liberalization of electricity markets, complementing RES-E directive's market-making measures, constituted a major Europeanization dynamic both in old and highly industrialized member states like Germany, the Netherlands, France or Italy and newer EU members such as Spain, Poland or Bulgaria.

However, it was primarily the processes of horizontal Europeanization, i.e. the direct diffusion or transfer of policies, models or ideas from one EU member state to another in the shadow of potential EU-wide harmonization, that positively influenced the instrumental design of many domestic RES-E policies. By setting a concrete and widely visible example for an effective support scheme, the proponents of FITs (Denmark, Germany and Spain) were able to influence the shape of RES-E policies in other member states (Busch and Jörgens 2012). In this context, the observation by Vogelpohl et al. (2017, 51) that 'Germany's support scheme served as a model for other countries and thus provided for horizontal Europeanization by learning and imitation processes' characterizes not only this phase of EU renewable energy policy, but also the negotiation and implementation of the 2009 RED. At the same time that the FIT system spread to countries like Italy, Poland took inspiration from the UK and adopted support schemes based on quotas and TGCs (Jankowska and Ancygier 2017, 188). Thus, rather than harmonizing national support schemes, horizontal Europeanization resulted in a continuous oscillation of support schemes between the more environmentally effective FITs and the more economically efficient TGCs or feed-in-premiums (FIPs).

Round 2: the 2009 RED and the deepening of integration

Conferring authority

By 2007, a review of the implementation of the RES-E directive showed the need for mandatory targets if the EU wanted to reach its climate mitigation goals (European Commission 2007). Taking advantage of the rising media and public attention to climate change (Solorio and Bocquillon 2017), in January 2008 the Commission proposed a climate and energy package which included the so-called '20-20-20' targets: a 20% reduction in GHG emissions (expandable to 30% in case of international agreement), a 20% energy saving target and a target of 20% RES in EU final energy consumption. In March 2007, a European Council 'entrapped' by previous commitments to act as an international leader on climate change (Skovgaard 2013, 1147) endorsed the 20-20-20 targets, including a binding target of a 20% share of RES in overall EU energy consumption by 2020. In January 2008, the Commission proposed a set of new legislative measures on climate and energy, which included a draft directive distributing the burden of 20% of RES among the member states. In the run-up to the Copenhagen Climate conference, the climate and energy package was object of a fast-track negotiation and by December 2008 it was agreed by the European Council (Wurzel and Connelly 2011, 8). The legitimacy that the European Council's endorsement gave to the RES targets kept the contestation on targets at a relatively low level (Solorio and Bocquillon 2017). But a renewed attempt led by the European Commission to harmonize national support schemes using TGCs did cause strong disagreements, requiring intense negotiations to break the deadlock (Lauber and Schenner 2011).

In 2009, Directive 2009/28/EC on RES promotion, also known as RED, replaced the RES-E directive (and also the biofuels directive which had targeted the transport sector). It established a common framework for RES promotion, including not only the electricity sector but also transport and heating and cooling, and established an overall RES target of 20% by 2020 as well as mandatory national targets. However, its implementation was left mainly in the hands of national governments, requiring them to elaborate National Action Plans with non-binding sub-sectoral and interim objectives for each sector. The Commission, in turn, was in charge of evaluating the action plans and the indicative trajectory towards national targets. In terms of 'market-making' measures, the directive established that member states should grant RES priority or guaranteed access to the grid system. Overall, the RED slightly increased the authority conferred on the Commission mainly due to the introduction of mandatory national targets (Toke 2008). But this was in practice limited by a decentralized policy framework that left implementation to the member states (Solorio and Bocquillon 2017, 29).

Sources of contestation and the management of authority conflicts

In the process that led to the RED adoption, the Commission started in 2007 a new attempt to promote an EU-wide model for support schemes based on TGCs (Lauber and Schenner 2011). Such an EU-wide support scheme for RES would have meant a significant extension of the EU's authority to determine the ways in which RES goals were to be reached nationally. In this context, the UK appeared once again as a defender of trade in RES and as an important ally of the Commission (Solorio and Fairbrass 2017, 110). By early January 2008 the circulating versions of the RED proposal were favourable to this position (Toke 2008, 3003). Despite their success in RES promotion, Germany and Spain were forced to veto an EU-wide harmonization of support schemes based on TGCs in order to preserve their domestic FITs (Vogelpohl et al. 2017, 51–52). Interestingly, the conflicts over support schemes were attenuated and compromise was made possible in part because the UK changed its negotiation strategy from one based on national preferences to one of coalition-building. In 2008, the UK together with Germany and Poland presented a joint proposal that discarded TGCs while introducing the concept of 'non-trading flexibility' (Lauber and Schenner 2011, 520). The acceptance of this proposal, that ended up being included in the RED, was also possible thanks to France's role as a 'honest broker' (Bocquillon and Evrard 2017, 170). This meant that while the binding prescription of an EU model for domestic support schemes was avoided, the directive facilitated the voluntary cooperation and coordination among member states via joint support schemes, joint projects and statistical RES transfers. Authority conflicts were again mitigated through the continuation of a soft governance approach that facilitated flexibility in the design and implementation of national support schemes.

In the same vein, the nature and level of ambition of the RES target as well as the breakdown of the overall target into specific sub-targets for the different types of RES was a particularly delicate issue because, for the first time, new member states – those that had entered the EU in the first and second round of Eastern enlargement – undertook serious attempts to upload their domestic policy preferences to the EU level (Jörgens, Öller, and Solorio 2017). Together with Italy, Eastern European member states were concerned about the economic costs of the Commission's proposal. Italy was opposed to national binding targets on the grounds of the economic challenge that they represented and because it was considered as 'an imposition to modify the energy mix' (Di Nucci and Russolillo 2017, 130). Representing various Eastern European member states, Poland also demanded a more flexible application of the climate and energy package (Jankowska and Ancygier 2017). On the other side, a group of traditional 'green' member states including Denmark and Germany backed the Commission proposal (Dyrhauge 2017, 95; Vogelpohl et al. 2017). In the middle were member states like the UK, France, the Netherlands and Spain which, although supportive to the RES goal, were more reluctant to translate it into binding national targets (Bürgin 2015, 696). Ultimately, the endorsement of the 20-20-20 targets at the European Council in March 2007 was considered an achievement of German Chancellor Angela Merkel's leadership (Solorio and Bocquillon 2017) combined with the pressure on some of the EU leaders that resulted from their previous climate commitments (Skovgaard 2013). A crucial part of the deal was the introduction of an economic justice criterion to the distribution of the RES target among the member states, the existence of non-binding sub-sectoral targets and the lack of explicit sanctions for non-compliance (Solorio and Bocquillon 2017). In fact, the most significant competence shifted to the Commission was the possibility to issue recommendations for member states on how best to achieve their national targets in case of non-compliance. Overall, national obligations were still considerably loose and the conferral of authority on the Commission ended up being much more limited than would have been the case with the original proposal (Solorio and Bocquillon 2017, 29–33).

Implementation and policy change

When assessing the RED implementation, direct top-down Europeanization plays an important role only in a relatively small number of countries, namely Italy and France as well as new member states such as Poland, Romania and Bulgaria (Jörgens, Öller, and Solorio 2017, 294–296). The main reason was that – apart from the mandatory national RES targets that were perceived as a great challenge for example in Italy, the Netherlands and France – the RED created only very limited direct adaptational pressure in the member states. Actually, it had its biggest impact on those member states that joined the EU after 2004 given that accession countries were under special scrutiny to comply with the entire body of EU secondary law (including those laws that were still in the making) (Davidescu 2017; Hiteva and Maltby 2017).

Concerning horizontal Europeanization, this phase was characterized by the parallel diffusion of two types of support schemes: the FITs, where Germany continued to be the European role model, and TGCs, with the UK being the main point of reference (Busch and Jörgens 2012). But horizontal Europeanization was not limited to the successful cross-national diffusion of support schemes, but also occurred at the level of specific settings. An example is the cross-national transfer of limits for installed photovoltaics (PV) capacity. Similar to what happened in Spain, annual limits on installed PV capacity were introduced in Italy by the mid-2010s (Di Nucci and Russolillo 2017). Finally, in cases where policy development at the national level was blocked or lagged behind the expectations of domestic proponents, the horizontal diffusion of ambitious policies sometimes shifted to the subnational level. For example, in the Netherlands proactive local governments drew inspiration for ambitious RES-E policies from their counterparts in Germany (Hoppe and van Bueren 2017).

Nevertheless, the implementation of the RED occurred in an unfavourable context. On the one hand, the setback at the 2009 Climate Summit in Copenhagen shed doubts on the alleged EU leadership in global climate politics and on the viability of a European energy transition (Wurzel and Connelly 2011). On the other hand, 'the financial crisis and its economic and budgetary consequences (…) fueled debates about the cost of RES support schemes, which have been blamed for rising electricity prices in several member states' (Solorio and Bocquillon 2017, 34). This context rebooted claims from part of some member states for the renationalization of the climate and energy governance framework (Bürgin 2015, 699).

Round 3: the 2018 RED II and the management of sovereignty surpluses

Conferring authority

In January 2014, the Commission put forward its proposal for a post-2020 climate and energy framework, including a 40% GHG emissions reduction target by 2030 combined with a 27% RES target and a similar target for energy efficiency (see Dupont forthcoming). Although binding at the EU level, no mandatory targets were proposed nationally 'in the name of flexibility' (Solorio and Bocquillon 2017, 35). This shift was the result of authority debates within the EU. On the one hand, the European Parliament and parts of the Commission were pressing for an ambitious binding RES target (Neslen 2014; Bürgin 2015). On the other, parts of the Commission and a bloc of member states led by the UK argued in favour of a technology neutral energy transition, showing mainly a substance-based contestation related to the means of achieving the decarbonization of the energy sector (Neslen 2014; EURACTIV 2014). In a context still marked by the economic and financial crises (Bürgin 2015, 698), the October 2014 European Council ended up endorsing the Commission's proposal. Only Denmark, Germany and the Netherlands pushed for a more ambitious binding RES target of 30% by 2030, while Spain and Italy backed a target of at least 27% (Dyrhauge 2017; Vogelpohl et al. 2017; Hoppe and van Bueren 2017). In the end, the 2030 climate and energy framework (with a 27% target for RES that is binding at the EU but not at the national level) suggested a trend towards the renationalization of the EU renewable energy policy's governance structure (Solorio and Bocquillon 2017, 34–36).

By the end of 2016, in a post-Paris Agreement context and with the purpose of implementing the 2030 climate and energy framework, the Commission put forward a package to speed up its energy transition in line with Commission President Juncker's Energy Union priorities. The package comprised measures such as a Regulation on the Governance of the Energy Union (see Bocquillon and Maltby forthcoming), the Energy Performance in Buildings Directive together with the Energy Efficiency Directive (see Dupont forthcoming), and the RED II. In relation to the latter, the text proposed an EU-wide target of 27% for 2030 and the opening up of national support schemes.

In February 2017, Ministers in the Energy Council underlined the need to make progress on all proposals and stressed the strategic relevance of the Energy Union. However, in relation to the RED II, member states remained divided over support schemes. While 'several ministers supported the move towards a more market-based approach' (European Council 2017, 8), in line with the guidelines on state aid for environ-mental protection and energy 2014–2020, many argued in favour of flexibility. On December 2017, the Council agreed to pursue the proposed 27% target. Yet, by January 2018, the European Parliament adopted a draft law envisioning a RES share of 35% by 2030 and, quite surprisingly, in April 2018 EU energy Ministers revealed that member states were reconsidering the ambition of RES targets (Morgan 2018). A trilateral agreement between negotiators of the Commission, the European Parliament and the Council was reached on June 2018, setting an EU-wide RES target of 32% by 2030 (including the same three sectors as the previous one) and containing a review clause by 2023 for an upward revision if necessary. This compromise was possible because, against all odds, in the final stretch of the RED II negotiations, a group of member states including Spain, Italy, Portugal, Germany, Austria, the Netherlands, Denmark and Luxembourg had accepted a more ambitious policy (Darby 2018b; Simon 2018a). A game-changer was the entrance of new governments in Spain and Italy, 'shifting the majority' (Simon 2018a) within the Council. Despite the pressures from the European Parliament and environmental campaigners, Germany's veto against a target above 32% of RES impeded higher ambitions (Simon 2018b).

The Commission remained in charge of assessing the member states' performance, but this time backed by the Regulation on the Governance of the Energy Union and Climate Action, a legislative piece that requires all member states to develop integrated National Energy and Climate Plans (NECPs) while giving the Commission the power to monitor national and EU progress towards achieving the energy and climate targets (see Bocquillon and Maltby forthcoming). Additionally, the opening of support schemes for producers located in other member states was approved and a clause on the stability of financial support was embraced. Overall, the final agreement increased the diffusion of authority in EU renewable energy policy.

Sources of contestation and the management of authority conflicts

The domestic impact of EU renewable energy policy is key for understanding the desire of member states to slow down European integration in this field. In the aftermath of the economic and financial crises, the contestation of EU authority came from concerns with high electricity prices, costly infrastructure investments and the competitiveness of domestic industries. This context even led countries such as Germany, Spain and Denmark, known for their success in RES promotion, to take a more reluctant approach towards the 2030 goals. The reasons for this shift are manifold, but several of our national case studies draw a picture of self-defeating success of RES-promotion aggravated by the economic and financial crises. For example, Europeanization in Bulgaria and Romania led to a fast and very effective transposition of EU directives that was later counteracted by non-compliance and a partial dismantling of the domestic RES-E support schemes (Jörgens, Öller, and Solorio 2017).

The financial and budgetary strains caused by a successful promotion of RES-E is by no means restricted to the less affluent members of the EU. A pioneer with respect to the dismantling of FITs was the Netherlands which abolished their successful support scheme in 2006, only three years after it had come into force (Hoppe and van Bueren 2017). A similar development can be observed in Spain whose very successful FITs underwent a stepwise dismantling in 2007/2008. Regarding France, by 2010 'the high level of the solar PV FITs was made responsible for a "speculative bubble" and rising electricity prices, and criticized for favouring technology imports over national industry support' (Bocquillon and Evrard 2017, 171). As in Spain, the PV FIT was eventually dismantled. Blame-shifting towards the EU was the defining feature of national government's performance, contributing to an environment of skepticism towards European integration amongst the population (Jörgens, Öller, and Solorio 2017).

In this context, the authority claims by member states were mainly driven by a substance-based contestation. The Poland-led Visegrad Group, representing Central European countries, pressed for greater flexibility and financial assistance to modernize their energy systems and meet future climate targets (Simon 2018b). For this group of countries, the problem was more about the purpose of EU renewable energy policy, which from their perspective should support modernization and economic development instead of representing a financial burden. Against this background, the UK appeared as the toughest opponent of the 32% of RES goal, calling along the negotiations for a 30% target (Vaughan 2018), and its commitment to EU goals after Brexit is uncertain (Darby 2018a).

The adoption of the 2018 RED II revealed two ways of settling authority conflicts. First, the abandonment of binding national targets placed limits on previously delegated authority without going as far as a full renationalization of the EU renewable energy policy. Second, diffuse authority was fine-tuned with the agreement on the monitoring of national performance. Despite the fact that member states are responsible for defining their national contributions to meet the collectively binding EU-target, the revision process is now regulated by the Energy Union Governance Regulation (Bocquillon and Maltby forthcoming). While national governments prepare integrated NECPs, i.e. ten-year period plans which must include specific RES goals, the Commission is responsible for assessing the draft plans and has the power to issue country-specific recommendations. On the one hand, this compromise, which has been defined as soft governance with harder edge (Oberthür 2019, cf. Bocquillon and Maltby forthcoming), settled concerns over the RES targets and the constraints it imposed upon the national government's control over their energy mix. On the other, part of the substance-based contestation over the purpose and costs of energy transition was solved thanks to a successful strategy of de-politicization. After several unsuccessful attempts to prescribe an EU model for support schemes, with the guidelines on state aid for environmental protection and energy 2014–2020, the Commission opted for the proscription of FITs as permissible state aid – having to be replaced, 'after a transitional phase' by tendering procures or TGCs (Vogelpohl et al. 2017, 53). Consequently, support systems were no longer the subject of heated discussions during the negotiations. Besides, the inclusion of RES targets within the energy governance regulation – with its emphasis on meeting the Paris Agreement – also contributed to reduce contestation.

Implementation

At the time of writing it is still early to assess the implementation of RED II. As a first step, member states were required to submit draft NECPs to the European Commission by the end of 2018. This deadline was missed by seven member states and a significant number of plans that were submitted on time did not follow the template provided by the European Commission and lack some of the required information (Morgan 2019). Finally, however, the 28 member states presented their first draft of NECPs, which have been subject to the Commission's scrutiny[2]. According to it (European Commission 2019a, 3), under current draft plans, the EU would fall below the 32% share of RES. By June 2019 the Vice-President for the Energy Union, Maroš Šefčovič insisted on the fact that final plans have to be ready by the end of 2019 and that by then the ambition gap has to be closed (European Commission 2019b). Apparently, this indicates that the EU's ambitious climate goals and their implications for the domestic renewable energy policies continue to be contested at the national level.

Conclusions

Based on the case of RES promotion, this article has dealt with authority debates in EU energy policy and their evolution over time. By employing an exploratory process tracing method and following three rounds of Europeanization, we have been able to system-atically analyze: (i) the delegation of authority to the EU; (ii) the contestation of EU authority by some of the member states as a reaction to issue-specific Europeanization processes; and (iii) the ways in which authority conflicts were managed during the negotiations of the RES-E directive, the RED, and the RED II. The Europeanization frame-work has proved to be a useful tool for exploring the complex causality behind EU's authority turns as well as for understanding the role played by procedural factors includ-ing recurrent instances of policy formulation and implementation and feedback loops between different rounds of policymaking. Our empirical findings call for a more intense use of Europeanization as an analytical path to examine the extent to which de-Europeanization is a consequence of feedback loops in EU policy-making.

In addition, our case study shows that the observation of feedback loops between different rounds of Europeanization is useful for exploring the complex causality behind the contestation of authority between the EU and its member states. A contestation that in some cases has even led to instances of de-Europeanization in the renewable electricity policy domain. Our case study also demonstrates that a research design that covers several rounds of Europeanization is a suitable strategy for examining the different ways in which authority conflicts are managed over time.

Regarding the framework of this special issue, this article has shown that, driven by a combination of functional needs (related to RES advantages for national energy sys-tems) and value-based objectives (related to the alleged EU international leadership on climate change), member states have tended, not always enthusiastically, to displace legal authority upwards in order to build and maintain a common EU renewable energy policy. Interestingly, given that this policy area touches upon national sovereignty over the energy mix, the delegation of authority to the EU has mainly followed the logic of soft modes of governance. Despite this, EU renewable energy policy has been able to produce changes nationally so that in recent years the policy has been characterized by a contestation by member states. While sovereignty-based contestation concerning national RES targets was solved via fine-tuning existing forms of soft governance in the context of the Energy Union's new governance regulation, substance-based contestation related to support schemes was managed with a (de)politization of the issue. Instead of pressing once again for a harmonization of support schemes, the Commission finally solved the issue, making use of the guidelines on state aid for environmental protection and energy 2014–2020. In this way, the analytical framework outlined in the introduction to this Special Issue demonstrates its usefulness for understanding authority debates in an EU immersed in a post-functionalist dilemma.

Notes

1. Since the 2009 RED directive the EU renewable energy policy covers three sectors (electricity, transport and heating and cooling). This article focuses exclusively on the electricity sector.
2. This information is available at the Commission webpage: https://ec.europa.eu/energy/en/topics/energy-strategy-and-energy-union/governance-energy-union/national-energy-climate-plans.

Acknowledgments

The authors would like to thank Jenny Fairbrass, Anna Herranz-Surrallés, Pierre Bocquillon, Maya Jegen and Aviel Verbruggen as well as the panel participants at the UACES Annual Conferences 2018 and 2019 in Bath and Lisbon and at the 4th International Conference on Public Policy 2019 in Montreal for valuable comments and suggestions. An anonymous reviewer is also thanked for critically revising the manuscript and suggesting substantial improvements.

Disclosure statement

No potential conflict of interest was reported by the authors.

ORCID

Helge Jörgens ⓘ http://orcid.org/0000-0002-0262-4189

References

Bocquillon, P., and A. Evrard. 2017. "Explaining the Uneven and Diffuse Europeanization of French Renewable Electricity and Biofuels Policies." In *A Guide to EU Renewable Energy Policy*, edited by I. Solorio and H. Jörgens, 162–182. Cheltenham: Edward Elgar.

Bocquillon, P., and T. Maltby. forthcoming. "EU Energy Policy Integration as Embedded Intergovernmentalism: The Case of Energy Union Governance Regulation." *Journal of European Integration* 42 (1).

Börzel, T. 2002. "Member State Responses to Europeanization." *Journal of Common Market Studies* 40 (2): 193–214. doi:10.1111/1468-5965.00351.

Bulmer, S., and S. Padgett. 2005. "Policy Transfer in the European Union: An Institutionalist Perspective." *British Journal of Political Science* 35 (1): 103–126. doi:10.1017/S0007123405000050.

Bürgin, A. 2015. "National Binding Renewable Energy Targets for 2020, but Not for 2030 Anymore." *Journal of European Public Policy* 22 (5): 690–707. doi:10.1080/13501763.2014.984747.

Busch, P.-O., and H. Jörgens. 2012. "Europeanization through Diffusion? Renewable Energy Policies and Alternative Sources for European Convergence." In *European Energy Policy*, edited by F. Morata and I. Solorio, 66–84. Cheltenham: Edward Elgar.

Coman, R. 2014. "*Concordia Discorse* from Cumulative Europeanization to Deeper Integration." In *Europeanization and European Integration*, edited by R. Coman, T. Kostera, and L. Tomini, 1–11. Basingstoke: Palgrave.

Copeland, P. 2016. "Europeanization and De-Europeanization in UK Employment Policy." *Public Administration* 94 (4): 1124–1139. doi:10.1111/padm.12283.

Darby, M. 2018a. "UK Government Refuses to Commit to EU Clean Energy Targets after Brexit." *Climate Home News*, June 6

Darby, M. 2018b. "EU Closes in on Clean Energy Package, with Spain, Italy Joining Push for Higher Targets." *BusinessGreen*, June 12.

Davidescu, S. 2017. "The Europeanization of Renewable Energy Policy in Romania." In *A Guide to EU Renewable Energy Policy*, edited by I. Solorio and H. Jörgens, 204–223. Cheltenham: Edward Elgar.

Di Nucci, M. R., and D. Russolillo. 2017. "The Fuzzy Europeanization of the Italian Renewable Energy Policy." In *A Guide to EU Renewable Energy Policy*, edited by I. Solorio and H. Jörgens, 121–140. Cheltenham: Edward Elgar.

Dupont, C. forthcoming. "Defusing Contested Authority: EU Energy Efficiency Policymaking." *Journal of European Integration* 42 (1).

Dyrhauge, H. 2017. "Denmark: A Wind-Powered Forerunner." In *A Guide to EU Renewable Energy Policy*, edited by I. Solorio and H. Jörgens, 85–103. Cheltenham: Edward Elgar.

EURACTIV. 2014. "UK, 'Czechs Call for Nuclear-friendly 2030 Energy Policy." *EURACTIV*, January 17.

European Commission. 2007. "Report on the Progress Made in the Use of Biofuels and Other Renewable Fuels in the Member States of the European Union (COM(2006) 845 final)." Brussels: European Commission. doi:10.1094/PDIS-91-4-0467B.

European Commission. 2019a. "United in Delivering the Energy Union and Climate Action - Setting the Foundations for a Successful Clean Energy Transition. COM(2019) 285 Final." Brussels: European Commission. Accessed 18 June 2019.

European Commission. 2019b. "Energy Union: Commission Calls on Member States to Step up Ambition in Plans to Implement Paris Agreement." Press release. Brussels: European Commission. June 17

European Council. 2017. Outcome of the Council Meeting, 3521 Council Meeting, Transport Telecommunications and Energy, Brussels, 27 February.

Exadaktylos, T., and C. M. Radaelli, eds. 2012. *Research Design in European Studies: Establishing Causality in Europeanization*. Basingstoke: Palgrave.

Herranz-Surrallés, A., I. Solorio, and J. Fairbrass. forthcoming. "Renegotiation of Authority in the Energy Union: A Framework for Analysis." *Journal of European Integration* 42 (1).

Hiteva, R., and T. Maltby. 2017. "Hitting the Target but Missing the Point: Failing and Succeeding in the Bulgarian Renewable Energy Sector." In *A Guide to EU Renewable Energy Policy*, edited by I. Solorio and H. Jörgens, 224–244. Cheltenham: Edward Elgar.

Hoppe, T., and E. van Bueren. 2017. "From Frontrunner to Laggard: The Netherlands and Europeanization in the Cases of RES-E and Biofuel Stimulation." In *A Guide to EU Renewable Energy Policy*, edited by I. Solorio and H. Jörgens, 65–84. Cheltenham: Edward Elgar.

Jankowska, K., and A. Ancygier. 2017. "Poland at the Renewable Energy Policy Crossroads: An Incongruent Europeanization?" In *A Guide to EU Renewable Energy Policy*, edited by I. Solorio and H. Jörgens, 183–203. Cheltenham: Edward Elgar.

Jörgens, H., E. Öller, and I. Solorio. 2017. "Conclusions: Patterns of Europeanization and Policy Change in the Renewable Energy Policy Domain." In *A Guide to EU Renewable Energy Policy*, edited by I. Solorio and H. Jörgens, 289–313. Cheltenham: Edward Elgar.

Lauber, V., and E. Schenner. 2011. "The Struggle over Support Schemes for Renewable Electricity in the European Union: A Discursive-institutionalist Analysis." *Environmental Politics* 20 (4): 508–527. doi:10.1080/09644016.2011.589578.

Liefferink, D., and M. S. Andersen. 1998. "Strategies pf the 'Green' Member States in EU Environmental Policy-making." *Journal of European Public Policy* 5 (2): 254–270. doi:10.1080/135017698343974.

Meyer, N. I. 2003. "European Schemes for Promoting Renewables in Liberalized Markets." *Energy Policy* 31: 665–676. doi:10.1016/S0301-4215(02)00151-9.

Morgan, S. 2018. "EU Member States Warm up to Parliament's Energy Stance." *EurActiv*, April 19.

Morgan, S. 2019. "Seven EU Nations Miss Climate and Energy Plan Deadline." *EurActiv*, January 11.

Neslen, A. 2014. "Parliament, Commission Set for Clash over 2030 Clean Energy Goals." *EURACTIV*, January 10.

Oberthür, S. 2019. "Hard or Soft Governance? the EU's Climate and Energy Policy Framework for 2030." *Politics and Governance* 7 (1): 17–27. doi:10.17645/pag.v7i1.1796.

Radaelli, C. M., and T. Exadaktylos. 2010. "New Directions in Europeanization Research." In *Research Agendas in EU Studies: Stalking the Elephant*, edited by M. Egan, N. Nugent, and W. E. Paterson, 189–215. Basingstoke: Palgrave.

Radaelli, C. M. 2018. "EU Policies and the Europeanization of Domestic Policymaking." In *Handbook of European Policies: Interpretive Approaches to the EU*, edited by H. Heinelt and S. Münch, 55–71. Cheltenham: Edward Elgar.

Reiche, D., and M. Bechberger. 2004. "Policy Differences in the Promotion of Renewable Energies in the EU Member States." *Energy Policy* 32: 843–849. doi:10.1016/S0301-4215(02)00343-9.

Rohlfing, I. 2012. *Case Studies and Causal Inference. An Integrative Framework*. Basingstoke: Palgrave McMillan.

Rowlands, I. H. 2005. "The European Directive on Renewable Electricity: Conflicts and Compromises." *Energy Policy* 33: 965–974. doi:10.1016/j.enpol.2003.10.019.

Saurugger, S. 2014. "Europeanisation in Times of Crisis." *Political Studies Review* 12 (2): 181–192. doi:10.1111/1478-9302.12052.

Saurugger, S., and C. M. Radaelli. 2008. "The Europeanization of Public Policies: Introduction." *Journal of Comparative Policy Analysis: Research and Practice* 10 (3): 213–219. doi:10.1080/13876980802276847.

Simon, F. 2018a. "Shifting Politics Offer Fresh Hope of EU Deal on Clean Energy Laws." *EurActiv*, January 8.

Simon, F. 2018b. "Germany Pours Cold Water on EU's Clean Energy Ambitions." *EurActiv*, June 11.

Skovgaard, J. 2013. "The Limits of Entrapment: The Negotiations on EU Reduction Targets, 2007–2011." *Journal of Common Market Studies* 51: 1141–1157. doi:10.1111/jcms.12069.

Solorio, I., and P. Bocquillon. 2017. "EU Renewable Energy Policy: A Brief Overview of Its History and Evolution." In *A Guide to EU Renewable Energy Policy*, edited by I. Solorio and H. Jörgens, 23–44. Cheltenham: Edward Elgar.

Solorio, I., and J. Fairbrass. 2017. "The UK and EU Renewable Energy Policy." In *A Guide to EU Renewable Energy Policy*, edited by I. Solorio and H. Jörgens, 104–120. Cheltenham: Edward Elgar.

Solorio, I., and H. Jörgens, eds. 2017. *A Guide to EU Renewable Energy Policy: Comparing Europeanization and Domestic Policy Change in EU Member States*. Cheltenham: Edward Elgar.

Toke, D. 2008. "The EU Renewable Directive-What Is the Fuss about Trading?" *Energy Policy* 36: 3001–3008. doi:10.1016/j.enpol.2008.04.008.

Tosun, J., and I. Solorio. 2011. "Exploring the Energy-Environment Relationship in the EU." *European Integration Online Papers* 15. Article 7

Treib, O., H. Bähr, and G. Falkner. 2007. "Modes of Governance: Towards a Conceptual Clarification." *Journal of European Public Policy* 14 (1): 1–20. doi:10.1080/135017606061071406.

Vaughan, A. 2018. "EU Raises Renewable Energy Targets to 32% by 2030." *The Guardian*, June 14.

Vogelpohl, T., D. Ohlhorst, M. Bechberger, and B. Hirschl. 2017. "German Renewable Energy Policy." In *A Guide to EU Renewable Energy Policy*, edited by I. Solorio and H. Jörgens, 45–64. Cheltenham: Edward Elgar.

Woll, C., and S. Jacquot. 2010. "Using Europe: Strategic Action in Multi-Level Politics." *Comparative European Politics* 8 (1): 110–126. doi:10.1057/cep.2010.7.

Wurzel, R. K. W., and J. Connelly, eds. 2011. *The European Union as a Leader in International Climate Change Politics*. Abingdon: Routledge.

Defusing contested authority: EU energy efficiency policymaking

Claire Dupont

ABSTRACT
EU energy efficiency policymaking has faced repeated contestation. Such contestation has been both *sovereignty-based*, when member states contest the EU's authority to make policy on energy efficiency (subsidiarity claims), and *substance-based*, when concerns are raised about the choice or ambition of a policy measure. Yet, energy efficiency has become one of the five dimensions of the Energy Union. It is thus an example of EU policymaking advancing even under contestation. I investigate three main strategies to manage these contestations: (1) framing and reframing energy efficiency to enhance and consolidate authority at EU-level and to mitigate contestations over this authority; (2) developing the legal framework; and (3) applying flexibility in policy measures and employing mixed soft and hard governance tools. These strategies are employed, and interact, over both a 'long game' and a 'short game'.

Introduction

With the entry into force of the Lisbon Treaty in 2009, energy efficiency officially became a central component of EU energy policy. The Energy Union brings together the variegated and overarching aims of EU energy policy (sustainability, competitiveness, security) and outlines energy efficiency as a means to achieve all three goals. Energy efficiency has been elevated to one of the five main dimensions of the Energy Union. It has even been further prioritised discursively under the frame 'energy efficiency first'. How did energy efficiency come to be so central to EU energy policy and to the Energy Union?

I analyse the gradual conferral of authority to the EU-level on energy efficiency policymaking. Energy efficiency policy measures cut across sectors, and include measures to save energy and improve energy efficiency in products, production processes, transport, buildings, energy production and transportation, and more. Energy efficiency 'policy' therefore cannot be considered a separate and distinct policy field: rather it is a means to achieve the objectives of energy policy more broadly (i.e. competitiveness, security and sustainability) across a range of sectors. In assessing the EU's energy efficiency 'policy', I focus on two major policy instruments – the Energy Performance of Buildings Directive

(EPBD) and the Energy Efficiency Directive (EED) – although several other measures and instruments exist.

I assess the contestation to EU policymaking efforts on energy efficiency, and how these contestations were managed. Contestations are *sovereignty-based*, when concerns over subsidiarity are raised, and *substance-based*, when concerns over the choice or ambition of a policy measure are raised. Drawing on the framework proposed by Herranz-Surrallés, Solorio and Fairbrass (2020, this issue), I delve into the management strategies employed by the EU (particularly by the European Commission). I find that several strategies have been employed, sometimes simultaneously, and that they play out over long and short time horizons.

EU energy efficiency policy measures have been proposed since the 1970s. Contestation has been a constant part of the policy development but has varied in intensity and form. In the 1970s and 1980s, when the EU had limited confirmed authority, contestation was also limited. From the 1990s onwards, sovereignty-based contestation became more prominent. By the mid-2000s, substance-based contestation was the main form of contestation. The research focus here is on how the EU has managed these varied forms of contestation.

The paper is structured as follows. In the next section, I outline the analytical framework, building on Herranz-Surrallés et al. (2020, this issue), and Oberthür (2019). Then, I trace the development of the conferral of authority to the EU-level over three time periods, highlighting the types of contestation that arose: the 1970s and 1980s; the 1990s and early 2000s; and since the end of the 2000s. Finally, I apply the analytical framework to understand how contestation was managed. I conclude that managing contestation has involved three main strategies: (1) framing and reframing energy efficiency to enhance and consolidate authority at EU-level and to mitigate contestations; (2) developing the legal framework (competence connection and eventually a specified energy competence); and (3) applying flexibility in policy measures and employing mixed soft and hard governance tools. These strategies are employed in both a 'long game' (the development of authority in the policy field over time) and a 'short game' (within a single policy cycle), with interactions between the long and the short games.

Understanding contestation in EU energy efficiency policy

In the analytical framework proposed by Herranz-Surrallés et al. (this issue), there are three conceptual phases in the renegotiation of authority in the Energy Union: (1) authority conferral; (2) authority contestation; and (3) managing authority conflicts. Furthermore, they suggest two types of contestation: *sovereignty-based* and *substance-based* contestation.

In the first step, analysing authority conferral implies mapping the authority patterns in the development of energy efficiency policy measures at EU level. The main players are the EU supranational institutions and the member states. From a legal perspective, the entry into force of the Lisbon Treaty in 2009 marked the conferral of authority on energy efficiency to the EU level. Article 194(1) of the Treaty on the Functioning of the European Union (TFEU) states: ' ... Union policy on energy shall aim, in a spirit of solidarity between Member States, to: ... (c) promote energy efficiency and energy saving ... '. However, lacking the legal competence to act in a particular policy domain has not always blocked

EU level action (Pollack 1994). Reasons for *de facto* conferral of authority through the creative connection of competences can be based on functional needs or on values, as outlined by Herranz-Surrallés et al. (this issue). I would add that the active pursuit of efforts to confirm the conferral of authority may also require entrepreneurial policy actors to highlight policy-linking (Maltby, 2013). In the Energy Union, the Commission and member states are therefore particularly important actors for understanding the conferral of authority.

In the second step, *sovereignty-based* and/or *substance-based* contestation of the (conferral of) authority can occur. I suggest two main means through which the EU's authority to make policy on energy efficiency is contested. First, sovereignty-based contestation manifests itself through arguments to uphold the subsidiarity principle (Article 5(3) TEU) – that policy should be made at the lowest level of governance possible. This claim argues against a functional logic/perceived functional need for supranational adoption of energy efficiency policy measures. Second, substance-based contestation occurs when actors contest the *choice or ambition* of measures. More sovereignty-based contestation may be expected when the process of conferring authority is ongoing or not yet confirmed, while substance-based contestation may be more likely once the conferral of authority is established. Sovereignty-based and substance-based contestation are also possible simultaneously, especially in policy areas that do not benefit from a unified vision among policy actors.

In the third step, contestation is managed. Here, Herranz-Surrallés et al. (this issue) suggest four main management strategies: formal adjudication and/or flexibility measures, in response to sovereignty-based contestation, and (de)politicisation and/or enhanced coordination in response to substance-based contestation. In cases with both sovereignty-based and substance-based contestation, multiple strategies may be employed. In addition, I suggest that framing can play a significant management role. A fine-tuned analysis of soft or hard governance responses to manage authority contestation may also reveal considerable nuance.

First, on framing, as Daviter states, 'policy frames identify what is at stake in an issue' (2011, 2). The perception by different policy actors of the importance of an issue can influence decisions. The success of any framing can result from several factors, including the nature of the issue, the nature of the actor, the institutional and political context, and broader external events (that could open up windows of opportunity for specific framings). When faced with both sovereignty-based and substance-based contestations, the EU institutions can employ framing and reframing as a management strategy in response to both. For example, a contestation based on subsidiarity concerns could be countered by a framing highlighting the supranational aspects of the issue. A contestation based on substance could be managed through a framing that emphasises multiple benefits of the proposed measures, or by framing the policy as a matter of urgency. As such, (de) politicisation can be part of the (re)framing process.

Second, employing soft or hard governance as a management strategy could be seen in a nuanced way. Soft and hard governance modes can be understood as a continuum. Sebastian Oberthür (2019) suggests four criteria to assess the degree to which governance modes are considered hard. These include: (1) formal status or bindingness; (2) the nature of the obligations; (3) prescriptiveness and precision; and (4) accountability and effective implementation. Elements of soft and hard governance modes even within single energy efficiency policy measures can be expected. This suggests that changes to governance modes – either towards harder or softer governance measures/modes – can be employed as a management strategy. This further implies that a simple categorisation of a policy measure as soft or hard is not straightforward: while some elements of a single policy may become 'harder' along this continuum, other elements may become 'softer'.

Finally, the three conceptual steps may occur simultaneously. For example, refining the legal framework and (re)framing can be seen as management strategies that occur *before* contestation occurs, during the process of authority conferral, and/or when confirming conferred authority. Framing can be regarded as a tool to confirm the conferral of authority *and* as a management tool that is used to avoid contestation (see below). Both this strategy and the refinement of the legal framework play out especially as long-term strategies, building on previous efforts to confer and confirm authority and to manage contestation. Reformulating policy measures along the soft/hard governance continuum, however, may play out over short time frames within a single policy cycle, in response to specific concerns.

In the following section, I trace the development of EU energy efficiency policymaking over time, discussing the conferral of authority to the EU and the types of contestation that arose.

Authority conferral and contestation in EU energy efficiency policy

Energy efficiency means reducing energy consumption, saving energy, using less energy for more output. Energy efficiency is not a goal in itself, but a means to achieve different aims. These aims can include improving energy security (lower reliance on foreign energy suppliers), improving energy sustainability (lower levels of fossil fuel-based energy consumption) and improving competitiveness (lower levels of consumption leading to lower costs).

The EU has adopted an array of policy measures to improve energy efficiency cutting across multiple policy sectors (see Table 1) and has agreed on overarching energy efficiency targets. The most recent of these is the indicative target agreed in 2018 to

Table 1. Selection of EU energy efficiency policy measures. Note: the table is by no means comprehensive, as measures in multiple policy fields also constitute energy efficiency policy measures (e.g. policies to improve fuel efficiency in cars).

Year	Legislation	Code
1989	Construction products Directive	89/106/EEC
1992	Energy labelling Directive	92/75/EEC
1992	Efficiency for hot-water boilers Directive	92/42/EEC
1993	SAVE Directive	93/76/EEC
2002	Energy performance of buildings Directive	2002/91/EC
2004	Promotion of cogeneration Directive	2004/8/EC
2005	Ecodesign Directive	2005/32/EC
2006	Energy services Directive	2006/32/EC
2009	Ecodesign Directive – recast	2009/125/EC
2010	Energy labelling Directive – recast	2010/30/EU
2010	Energy performance of buildings Directive – recast	2010/31/EU
2012	Energy efficiency Directive	2012/27/EU
2018	Energy Performance of Buildings Directive	EU 2018/844
2018	Directive on energy efficiency	EU 2018/2002

Source: own compilation.

achieve 32.5 per cent improvement in energy efficiency by 2030, compared to 2007 models of business as usual projections.

Energy efficiency policy development has not gone uncontested. I trace the development of EU energy efficiency policy in three broad phases: the 1970s and 1980s; the 1990s and early 2000s; and, since the end of the 2000s.

1970s – 1980s: first, tentative steps

At this time, there was limited development on the EU-level on energy efficiency. Only very weak policy measures were agreed, and there was limited authority conferral to the EU-level. There were also few contestations over the EU's authority. The EU did not have competence to make policy on energy efficiency, but it could agree to some policy measures by linking to other policy areas. Policy discussions during these decades paved the way for future expansions of authority.

Energy efficiency came onto the EU's agenda in the wake of the 1970s oil crises. It was prominently framed as a logical response to energy security concerns. Improving energy efficiency, and reducing demand for oil, was seen as an effective measure to reign in growing dependence on foreign energy imports (Boasson and Dupont 2015; Boasson and Wettestad 2013; European Commission 1979a, 1984). Several member states adopted energy efficiency policy measures. National level energy performance standards for new buildings were adopted in the Netherlands, Denmark, Germany and France (European Commission 1979b, 2). Coordinated EU action started to emerge, but was hardly effective. This action was rather an indication of the type of EU policy measures that would eventually emerge. The Commission's 1979 report on the EU's early programme for energy saving highlighted that 'future savings will increasingly require investment in new equipment or buildings, or retrofitting the old, and more energy conscious behaviour from both investors and consumers' (European Commission 1979b, 1).

In 1986, the Council set the objective of improving energy efficiency in the EU by 20 per cent by 1995 compared to 1985 levels (Council of the European Union 1986) but this target was not met. The Council's 1986 resolution emphasised the same framing of energy efficiency as a response to energy security. Contestation towards efforts to make policy on energy efficiency at the EU level were not yet very vocal.

1990s – 2000s: expanding policy and contestation

In the 1990s and 2000s, EU energy efficiency policymaking gained momentum. While legal authority was still not transferred, there was a clear competence connection with internal market and environment policy domains. This allowed for some transfer of authority. Contestation over the EU's authority correspondingly increased, with both sovereignty-based contestations and substance-based contestations being raised. Calls to respect the subsidiarity principle echoed throughout policy negotiations, and regularly led to weakened and poorly implemented policy measures (Boasson and Wettestad 2013; Dupont 2016; Henningsen 2011). As the 2000s progressed, there were fewer *sovereignty-based contestations*, and more *substance-based contestations* on the choice and ambition of policy measures.

Energy efficiency was reframed by the Commission (and Council) in response to the adoption of the 1992 UN Framework Convention on Climate Change (UNFCCC) and its Kyoto Protocol (1997) as a measure to combat climate change (European Commission 1992, 1998). As climate change moved up the political agenda, energy efficiency became entangled in the EU's response (Oberthür and Roche Kelly 2008). Boasson and Wettestad outline that the international climate negotiations provided an opportunity for the Commission to push for ambitious energy efficiency measures in the EU (2013, 152). However, internal policies were largely ineffective (Oberthür and Roche Kelly 2008). One of the EU's attempts at making broad policy to promote energy efficiency – the SAVE programme in the 1990s – was substantially weakened during the policy process, with its budget dramatically reduced (Dupont 2016). When sovereignty-based contestations did not succeed in blocking policy adoption or in repatriating authority, substance-based contestations over the *choice* and *ambition* of policy measures led to weak policy output and generally succeeded.

The Energy Performance of Buildings Directive (EPBD) faced considerable and repeated contestation. The 2001 EPBD proposal cites climate change as a major rationale for taking action on improving the energy performance of buildings (European Commission 2001, pp. 15, 19). The Directive aimed to set out minimum standards for the energy performance of buildings but was applicable only to new buildings and to large existing buildings that were to undergo renovations, meaning that it had little effect (Henningsen 2011).

During the EBPD negotiations, two main recurring concerns surfaced: (1) from the perspective of subsidiarity, or a *sovereignty-based contestation*; and (2) about the cost of the policy measures, or a *substance-based contestation*. Objections about the need for such a policy measure on subsidiarity grounds were raised within the Commission itself, by three Commissioners: Chris Patten, then Commissioner for external relations, Neil Kinnock, then Vice President of the Commission, and Frits Bolkestein, then Commissioner for the internal market and services (Dupont 2016). Their objections led to redrafting and delays in the publication of the proposal. Member states in the Council were concerned about the application of the policy measures within their own territory, which motivated their push for delayed implementation and heightened flexibility (ibid.). The European Parliament questioned the cost-effectiveness of the proposed measures, and the Council inserted additional flexibility to reduce costs, including limiting the implementation of minimum performance standards to 'major' renovations and allowing a longer delay for implementation. The final Directive was heavily criticised by stakeholders for how much it had been weakened in the policy process. Blame was levelled against member states and against MEPs for capitulating to states' demands (ENDS Europe 2001, 2002). The Directive had little impact on energy performance overall (Henningsen 2011), and the Commission launched its preparations for a recast proposal even before the implementation phase of the EPBD (ENDS Europe 2005).

In 2007, the EU adopted its indicative target to improve energy efficiency by 20 per cent (compared to business-as-usual projections) by 2020 as part of its contribution to the UNFCCC Copenhagen Climate Summit in 2009. It also agreed on a binding target to reduce greenhouse gas emissions by 20 per cent, and to increase the share of renewable energy to 20 per cent by 2020 (European Council 2007, Solorio and Jörgens 2020).

The Commission published a proposal for a recast of the EPBD in 2008 (European Commission 2008a). The proposal was partly in response to the 2020 energy efficiency target. Negotiations on the EPBD recast took place alongside the negotiations on the measures in the 2020 climate and energy package that included proposals on the Emissions Trading System, renewable energy, effort-sharing of emissions reductions not covered by the Emissions Trading System and carbon capture and storage (Boasson and Wettestad 2013; Oberthür and Pallemaerts 2010). While agreement on the climate and energy package was reached in December 2008 under the French Presidency, negotiations on the EPBD took several months longer, with agreement achieved in 2009. The motivation to agree in time for the international climate negotiations in Copenhagen pushed the Swedish Presidency (in 2009) to ensure a compromise agreement was reached (Dupont 2016).

The recast Directive was considered necessary to fill the gap in implementation of the first EPBD and also to 'extend the scope, simplify its implementation and develop energy performance of buildings certificates into a real market instrument' (European Commission 2008b, 11). The most important adjustment was that *all* buildings undergoing major renovations would have to meet minimum energy performance standards. It also set out the benefit for member states of developing national policies for the greater uptake of low- and zero-energy buildings (European Commission 2008a, 6).

There were far fewer contestations based on subsidiarity claims (sovereignty-based contestation). A coalition of NGOs and business actors came together with the Parliament and the Commission to push forward on energy efficiency measures – there were few stakeholders who argued against action on energy efficiency (interview 1). However, substance-based contestations came from member states arguing in favour of taking account of cost considerations and flexibility in implementation.

This time, the Parliament was more ambitious and pushed back on member states. Its amendments included higher ambition for moving to zero-energy buildings; a shorter deadline for achieving a zero-energy ambition for all new buildings ; and an energy efficiency fund. Member states were not keen on the level of ambition of the Parliament (ENDS Europe 2009b, 2009c, 2009a), but external stakeholders were pleased. One member of an environmental NGO described the Parliament's first reading report as the 'best first reading ever' (interview 1). Nevertheless, Council watered down the Parliament's ambition to 'nearly' zero-energy (without clearly defining this) for new buildings by 2020 and the Directive provided few incentives for renovating the building stock (Boasson and Dupont 2015). The flexibility was again motivated by concerns over costs, and questions about the need for stringent and detailed measures at EU-level, emphasising the need to take into account differing national contexts (Boasson and Wettestad 2013; ENDS Europe 2009c).

Since the end of the 2000s: established legal framework

In 2009, an established legal framework confirmed the authority of the EU to make policy on energy efficiency, with the entry into force of the Lisbon Treaty (agreed already in 2007). This did not necessarily put an end to contestation, but it emphasised the shift in type of contestation from mainly *sovereignty-based* to mainly *substance-based*.

New frames to promote EU energy efficiency policy also began to emerge: 'win-win' arguments highlighting 'multiple benefits' of energy efficiency measures were emphasised (Fawcett and Killip 2019), including for jobs creation and cost competitiveness. After the disappointment of the climate negotiations in Copenhagen in 2009 (Dimitrov 2010), and given the depth of the financial and economic crises, political momentum waned. The broader context of multiple converging crises (Falkner 2016) drew attention away from energy efficiency. Energy efficiency did not enjoy the same status as central climate and energy files, and interest in it as a(n) (cost-)effective policy tool also faltered. This context made the negotiations leading to the 2012 Energy Efficiency Directive (EED) complex and conflictual, particularly as a result of substance-based contestation (Boasson and Dupont 2015).

The Commission put forward its proposal for the EED on 22 June 2011 (European Commission 2011) in light of continuing disappointment with energy efficiency policy measures, and with fears that the 2020 indicative energy efficiency target would be missed (ENDS Europe 2010). The EED repealed the Directive on energy end-use and energy services, and the Directive on cogeneration of heat and electricity, and aimed to fill the gaps left by the recast of the EPBD. It attempted to streamline energy efficiency policy measures and encourage proper implementation.

Negotiations centred around three main sticking points. First, the Parliament called for binding national targets to improve energy efficiency, while member states were opposed to this. Second, a renovation target of 3 per cent per year for all public buildings was supported by the Parliament, while member states aimed to reduce the remit of buildings covered. Third, a 1.5 per cent annual energy savings obligation for energy suppliers was also a point of controversy. Member states wished to provide flexibility for energy suppliers, including by staggering implementation over several years, or by including past efforts towards the target (ENDS Europe 2012a, 2012b). The job-creating potential was emphasised by lobbyists and policymakers in favour of the EED, in the context of a prolonged economic crisis.

Contestation around the EED stemmed predominantly from member states. Since the entry into force of the Lisbon Treaty, national parliaments have the opportunity to scrutinise policy proposals. For the EED negotiations, nine national parliamentary chambers from eight countries submitted contributions.[1] None argued against the EED on subsidiarity grounds (no sovereignty-based contestation), but the majority argued in favour of flexibility in implementation. They urged for consideration of the cost and administrative burdens the EED may add (substance-based contestation). The Italian Chamber, for example, suggested that there should be 'a verification of the practicability of requiring each Member State to carry out, within the time limits set, a detailed survey of public buildings' (Italian Chamber 2011). The Czech Senate recommended that 'the design and level of commitment' should be left in the hands of Member States because 'setting own country specific procedures at the national level will be more appropriate to the challenges each country is facing' (Czech Senate 2011). The Czech Senate argued against the adoption of stringent measures, and highlighted the added administrative and financial burden.

A compromise was found after trilogue negotiations. Member states were required to set 'indicative national energy efficiency targets', schemes and programmes (Recital 13, Art. 3.1), as their contribution to the EU's target to improve energy efficiency by

20 per cent compared to business as usual scenarios for 2020. Public building renovation should take place at the rate of 3 per cent per year, but only central government buildings fell within the scope (Art. 5). Interestingly, governments were required to set down renovation roadmaps for their entire stock of buildings (Art. 4). Energy sales to customers had to achieve 1.5 per cent savings per year to 2020 (Art. 7), but a number of flexible measures count towards this target. In sum, while there still were no binding national targets agreed under the EED, there were several far-reaching policy measures. Contestations were focused on avoiding a rigid and far-reaching Directive that was seen to add administrative and financial burdens.

But this was not the end of the EED. In October 2014, the European Council adopted climate and energy targets to 2030, including an indicative target to achieve at least 27 per cent improvement in energy efficiency – a target that was revised upwards to 32.5 per cent in 2018 (European Council 2014). In 2015, the Commission published its Communication on the Energy Union, where energy efficiency featured as one of five dimensions (European Commission 2015, 4). Energy efficiency had been elevated in policy importance to become a dimension of the Energy Union in its own right. With the adoption of the five dimensions of the Energy Union, the Commission was careful not to assign a hierarchy of importance to one dimension over another. This was a deliberate choice (interview 3), which has since evolved towards an 'energy efficiency first' framing (Bayer 2018).

Agreeing on the implementing measures (under the Energy Union) that aim to achieve the 2030 climate and energy goals required a long period of interinstitutional negotiations. The Commission launched its proposals in 2016, with 'putting energy efficiency first' as one of its three main goals. The Commission reframed energy efficiency as a single solution for multiple energy policy goals, but also for the overarching political agenda of creating new jobs, helping tackle (energy) poverty and improving air quality in Europe. The clarion call in support of energy efficiency was taken up by the European Parliament and civil society groups (interview 4), to eventually lead to a ratcheting up of the 2030 goals, and an agreement on the revised Energy Performance of Buildings Directive and revised Directive on Energy Efficiency (see Table 1).

The revised 2018 EED and EPBD strengthen policy measures in place or speed up/ improve implementation. The EPBD aims to update measures for national long-term building renovation strategies, to accelerate the rate of building renovations, and to support 'smart' solutions. The EED sets out the overarching target of achieving 32.5 per cent improvement in energy efficiency by 2030 and includes an option that this target could be ratcheted up further. Importantly, the Energy Union's Governance Regulation (EU 2018/1999, see Bocquillon and Maltby 2020) was also adopted in 2018. This piece of legislation requires member states to draw up National Energy and Climate Plans (NECPs) outlining their contributions and how they intend to meet their targets. By the end of the 2010s, we can see that energy efficiency policy measures have become more firmly embedded in the EU's policymaking sphere.

Negotiations on the revised EED were divisive. In contrast, negotiations on the revised EPBD went smoothly, with an initial agreement reached in Parliament in October 2017 (Morgan 2017). The main challenges in the EED negotiations were substance-based. First, there was considerable divergence among MEPs and member states on the nature and level of ambition of the overarching energy efficiency target, which was relevant both for the EPBD and EED. While MEPs called for a binding, ambitious target, member states supported a non-binding target. Trilogue negotiations (under the Bulgarian Presidency in 2018) faltered on this issue: when MEP Miroslav Poche, who was leading the negotiations on behalf of the Parliament proposed a non-binding target of 33 per cent, the Bulgarian Presidency could not agree. Eventually, a compromise agreement was found for a non-binding 32.5 per cent improvement in energy efficiency for 2030 (Keating 2018), based on 2007 modelled scenarios of business as usual energy consumption.

Second, negotiations were difficult on the specific measures. While member states emphasised the role of energy efficiency measures in final energy consumption, MEPs and civil society actors underlined the role of energy savings in primary energy consumption (making savings during electricity production also) (Simon 2018). The final formulation in the Directive provides for both options, but seems to leave room for flexibility that the emphasis may remain primarily on final energy consumption: 'Member States should set their national indicative energy efficiency contributions taking into account that the Union's 2030 energy consumption has to be no more than 1 273 Mtoe of primary energy *and/or* no more than 956 Mtoe of final energy' (Directive 2018/2002, Recital 6, emphasis added).

In the European Commission's responses to the member states' draft NECPs, it recommended that 20 member states should increase or substantially increase their ambition on energy efficiency.[2] Of the remaining member states, Austria, France, Germany, the Netherlands and Poland were invited to review their level of ambition in light of the EU 2030 goals; Luxembourg was requested to define and clarify its contribution; and Italy and Spain were recommended to explore the need for further implementation measures. Emphasising efficiency measures in both primary and final energy consumption was also a recurring theme in many of the recommendations. For 19 member states, the European Commission specifically underlined the need for clarity or heightened ambition on energy efficiency measures for primary energy consumption.[3]

Managing contestation

Having analysed the development of EU energy efficiency policymaking over time, I highlight two main developments regarding the contestation of authority. First, contestation has been a constant part of policy development, but has, second, varied in terms of intensity and type. In the 1970s, when the EU had little authority, there was a corresponding low intensity of contestation. From the 1990s onwards, contestation intensified. Evidence for both sovereignty-based contestation – by calling for the principle of subsidiarity to be upheld – and substance-based contestation – on the choice or ambition of the policy measure – was found. However, sovereignty-based contestation was dominant in the 1990s and early 2000s. Since then, contestation has since been *mostly* substance-based.

Management strategies can be divided between, first, a 'long-term game' of building discourse, authority and practice around energy efficiency policy development and, second, a 'short-term game' of dealing with contestation surrounding an energy efficiency policy instrument, with considerable interaction between them. I discuss three main management strategies: (1) framing and reframing energy efficiency; (2) developing the legal framework; and (3) applying flexibility in policy measures and employing mixed soft and hard governance tools. I distinguish between the first two strategies that play out over what I call a 'long game', namely the legal framework and framing, and the third strategy that is more likely to be part of a 'short game' within a policy cycle, namely agreeing flexible mixes of soft and hard governance measures. This third strategy also plays into the 'long game' as policy choices from one policy cycle to the next build upon each other.

The long game: framing and the legal framework

The conferral of authority to the EU level has taken time. There have been efforts to frame or reframe energy efficiency as a supranational issue, connected to internal market competences, to energy security concerns and to environmental issues. Energy efficiency policy measures advanced even without competence to act on energy issues, but the *de facto* competence creep (Pollack, 1994) was followed by Treaty change. Authority was conferred through the acceptance of EU level policy measures and was confirmed when the Lisbon Treaty was adopted (in 2007) and entered into force (in 2009). With the Treaty, the EU gained legal competence to make policy on energy efficiency under Article 194(1), while political agreement on this conferral of authority came earlier. Framing has thus been a strong part of the process of authority conferral and in reinforcing the moves towards a clear and established legal framework.

Framing and reframing has also been part of the 'long game' to avoid or overcome contestation, as seen also in the discussion on the historical development above. This includes consolidating authority to make policy on energy efficiency at the EU level and to avoid or limit contestation. Such (re)framing efforts could be seen as pre-emptive management strategies. Within a single policy cycle, the framing takes place before a policymaking process gets under way; responds to the wider context; and sets the tone for the ensuing negotiations. In the broader development of the policy domain, framing plays out over the long term to enhance or confirm authority and to prepare for, and justify, future policy developments by limiting room for (acceptable) contestation. By framing energy efficiency measures in a certain way (e.g. by linking them to climate change efforts and the EU's global role), contestation over the reasoning behind such policy proposals can be mitigated. The framing process can also be part of a strategy to (de)politicise energy efficiency, depending on whether the context calls for high politics or low politics solutions (Dupont 2019).

These 'long game' strategies seem most suited to respond to sovereignty-based contestation. However, even now that authority has been largely conferred to the EU level, and that sovereignty-based contestation has waned, a new 'energy efficiency first' framing has been put forward. This could be a strategy to keep contestation at bay, or also a strategy to manage substance-based contestation (to ensure policy outputs are strong and well implemented). The long game and the short game thus play out together.

The short game: flexibility in governance modes

The 'short game' plays out predominantly within a single policy cycle and has resulted in considerable flexibility in the ambition and choice of policy measures. However, the short game can have consequences over the long term, by establishing a degree of path dependency of choices for policy measures. There seems to be a long-term trend towards what can be considered hard governance modes, through the short-term negotiation on governance modes within a policy cycle.

In 1980s and 1990s, policy measures were rather declaratory, and programmes to support energy efficiency were so weakened in the policy process that they had little to no practical effect (Henningsen 2011). In the 2000s, the first EPBD was so flexible in its measures, that its recast was proposed *before* it was implemented as it already was clear how insufficient it was. The targets for improvements in energy efficiency to 2020 and to 2030 are indicative and non-binding.

'Short game' flexibilities within a policy cycle, can be seen as management strategies to deal with *substance-based contestation*, such as on the scope; choice; cost; or ambition of policy measures. These strategies also demonstrate a certain fluidity in the governance modes applied (hard or soft) and that the policy cycles build upon each other, (contributing to the development of energy efficiency policy measures in general in the 'long game').

Delving deeper into the strategy to use soft or hard governance modes to manage contestation, reveals a mixed story for energy efficiency. By applying Oberthür's (2019) four criteria of hard/soft governance, I find that energy efficiency has traits of both hard and soft governance – but has been moving towards more hard governance modes. These criteria are: (1) formal status or bindingness; (2) the nature of the obligations; (3) prescriptiveness and precision; and (4) accountability and effective implementation.

First, energy efficiency remains 'soft' from the perspective of the formal status or bindingness of the overarching goals. The goal to achieve energy efficiency improvements of 20 per cent by 2020 was indicative. The 2030 goal of 32.5 per cent (negotiated upwards from 27 per cent) improvement in energy efficiency is also indicative. Nevertheless, the possibility of missing the 2020 indiciative target was part of the motivation to propose new policy measures at EU level (e.g. the 2012 EED) (ENDS Europe 2010; European Commission 2011). It started to become clear that if measures were not adopted at the EU level, action on energy efficiency was not guaranteed at member state level (interview 2). The final output of the 2018 EED negotiations also includes the possibility to increase the 2030 indicative target further. The existence of the indicative target – more of a 'soft' governance approach – has nevertheless allowed for the consolidation of authority at EU-level.

Second, certain policy elements (or policy obligations) adopted must be implemented, and infringement procedures are possible if member states do not correctly do so. The 2010 EPBD, for example, set standards for the energy performance of new buildings and renovations. The 2012 EED included measures obliging member states to develop renovation roadmaps, and energy sales to customers have to achieve 1.5 per cent savings per year to 2020. Under the 2018 Energy Union Governance Regulation, member states are obliged to submit National Climate and Energy Plans. These are examples of how *elements* of policy measures or specific *policy obligations* can be binding, even while the overarching policy target remains indicative – thus combining hard governance tools with soft overarching governance modes.

Third, the precision and prescriptiveness of energy efficiency policy measures have been refined through successive policy cycles. For example, the 2002 EPBD set out standards for new buildings only. The 2010 EPBD refined this to include standards for renovations. However, the standard to achieve 'nearly-zero energy' buildings remained undefined (low precision). The 2012 EED specified that renovating public buildings should occur at a rate of 3 per cent per year, which clarified the EPBD provision that public buildings should be 'nearly-zero energy' by 2020. The 2018 EPBD aims to accelerate renovation further with long-term renovation plans in member states. Flexibility in implementing energy efficiency measures persists, but there has been increased precision and prescriptiveness of the policy obligations – indicating a slow move towards hard governance tools.

Fourth, accountability and effective implementation have also improved with each policy cycle, albeit slowly and with occasional setbacks. Implementation has been a particular challenge. Weakening policy ambition and/or implementation measures during negotiations has been part of the problem – too much flexibility in measures or in the timeframe for implementation has led to few effective outcomes (Dupont 2016). The 2002 EPBD, for example, had no discernible effect on the improvement of energy efficiency (Henningsen 2011). Energy efficiency has improved in the EU. In 2016, primary energy consumption was 10 per cent lower than it was in 2005 (EEA 2018). But it is very difficult to establish how much (if any) of this improvement is causally linked to EU policy measures (instead of to economic downturn, for example). On accountability, it has become easier for the Commission, and civil society organisations, to hold member states to account for their (in)action. The Commission oversight of the NECPs is one important example of an accountability mechanism (Bayer 2018; Oberthür 2019). This can again be seen as a hardening of governance modes.

In sum, while overarching targets for improving energy efficiency have remained non-binding, policy measures have moved towards hardening governance tools over successive policy cycles. Ensuring flexibility, while using elements of harder governance modes, is a management strategy.

The long and the short of it

Given the multiple strategies to mitigate and manage both sovereignty-based and substance-based contestation in energy efficiency policymaking, it is important to examine the long and the short games together. While framing and the development of the legal framework were strategies that seemed most suited to respond to sovereignty-based contestation, this 'long game' also affected the 'short game', and *vice versa*. While flexibility has been key to the management of substance-based contestation, allowing the adoption of specific policy instruments on energy efficiency, we nevertheless see a trend towards a hardening of energy efficiency governance. This implies a link between the management strategies: framing and reframing in the long game and in pre-emption of, or in response to, sovereignty-based contestation interacts with the ability to manage substance-based contestation. Clearly, the short game within a policy cycle is also part of the long game: decisions within policy cycles build upon each other, leading to room for further confirmation of authority. A more detailed study would be required to test the interconnections, but the energy efficiency case reveals that multiple management strategies can be employed simultaneously and/or over different timescales. Together, these strategies contribute to managing both types of authority contestation.

Conclusion

EU policymaking on energy efficiency is a fascinating example of how the EU suprana-tional level has framed, justified, pre-empted and managed different types of constesta-tion over different timescales using different management strategies, sometimes deployed simultaneously. Through this analysis, this paper contributes to deepening understanding of both the conferral of authority and the management of authority contestations in the EU.

EU policy has been developing since the 1970s. Authority was not legally conferred to the EU until the Lisbon Treaty, which entered into force in 2009. But that did not prevent the EU from negotiating and adopting measures on energy efficiency: energy efficiency was framed as a response to energy security converns and climate change, and measures were linked to the EU's internal market or environment competences. The authority of the EU to make policy on energy efficiency was eventually confirmed, in practice and in the legal framework.

However, policy proposals did not go uncontested. Contestation has been a constant part of policy development, but has varied in terms of its intensity and its type. In the 1970s and 1980s, when the EU had limited authority, contestation was also limited. Later, from the 1990s onwards, sovereignty-based contestation became more prominent. By the mid-2000s, substance-based contestation was the main form of contestation.

Managing contestation involves three main strategies: (1) framing and reframing energy efficiency to enhance and consolidate authority at EU-level and to avoid and mitigate contestations; (2) developing the legal framework (competence connection and eventually a specified energy competence); and (3) applying flexibility in policy measures and employing mixed soft and hard governance tools. Strategies (1) and (2) seem more suited to managing sovereignty-based contestation over a long time period (the devel-opment of the policy field). Strategy (3) seems more suited to managing substance-based contestation in negotiations on a particular policy instrument over a short time period (within a single policy cycle). There are interactions between the 'long game' and the 'short game', including the framing process within a policy cycle being influenced by the long game, and the successive short games building into the development of the long game. In energy efficiency, there is a trend towards 'harder' governance measures over successive policy cycles, meaning that the third strategy could represent an important, general feature of energy efficiency policy development at the EU level. These interactions merit further research, and establishing how these strategies are employed and interact across policy domains could prove a valuable avenue for future research.

Notes

1. The Czech Senate, the Italian Senate, the Portuguese Parliament, the Austrian Bundesrat, the German Bundesrat, the Irish House of the Oireachtas, the Italian Chamber, the Luxemburgish Chamber and the Romanian Chamber.
2. European Commission recommendations on the draft NECPs, available at: https://ec.europa.eu/energy/en/topics/energy-strategy-and-energy-union/governance-energy-union/national-energy-climate-plans, last accessed: 13 November 2019.
3. Austria, Bulgaria, Croatia, Cyprus, Czechia, Denmark, Estonia, Finland, Germany, Greece, Hungary, Ireland, Latvia, Lithuania, Luxembourg, Malta, Romania, Slovakia and Slovenia.

Acknowledgments

The author would like to thank the journal issue editors and the anonymous reviewer for very helpful and insightful feedback.

Interviews

Interview 1: NGO representative	10 January 2012
Interview 2: DG Energy Official	20 March 2013
Interview 3: DG Climate Action Official	10 March 2016
Interview 4: NGO representative	19 April 2018

Disclosure statement

No potential conflict of interest was reported by the author.

References

Bayer, E. 2018. *Energy Efficiency First: A Key Principle for Energy Union Governance*, 1–8. RAP. https://www.raponline.org/wp-content/uploads/2018/04/rap-bayer-key-principle-for-energy-union-governance-2018-april-17.pdf

Boasson, E. L., and C. Dupont. 2015. "Buildings: Good Intentions Unfulfilled." In *Decarbonization in the European Union: Internal Policies and External Strategies*, edited by C. Dupont and S. Oberthür, 137–158. Houndmills: Palgrave MacMillan.

Boasson, E. L., and J. Wettestad. 2013. *EU Climate Policy: Industry, Policy Innovation and External Environment*. Farnham: Ashgate.

Bocquillon, P., and T. Maltby 2020. "EU Energy Policy Integration as Embedded Intergovernmentalism: The Case of Energy Union Governance Regulation." *Journal of European Integration* 42 (1), forthcoming.

Council of the European Union. 1986. "Resolution Concerning New Community Energy Policy Objectives for 1995 and Convergence of the Policies of the Member States." *15/16.IX.86*, (Annex I).

Czech Senate. 2011. *463rd Resolution of the Senate on the Proposal for a Directive of the European Parliament and of the Council on Energy Efficiency*. Prague: Czech Senate.

Daviter, F. 2011. *Palgrave Studies in European Union Politics*. Houndsmills: Palgrave MacMillan.

Dimitrov, R. S. 2010. "Inside Copenhagen: The State of Climate Governance." *Global Environmental Politics* 10 (2): 18–24. doi:10.1162/glep.2010.10.2.18.

Dupont, C. 2016. *Climate Policy Integration into EU Energy Policy: Progress and Prospects*. London: Routledge.

Dupont, C. 2019. "The Collective Securitization of Climate Change." *West European Politics* 42 (2): 369–390. doi:10.1080/01402382.2018.1510199.

EEA. 2018. *Annual Indicator Report Series (AIRS): Energy Efficiency*. Copenhagen: European Environment Agency.

ENDS Europe. 2001. "Ministers Neuter EU Building Efficiency Drive." December 5.

ENDS Europe. 2002. "MEPs Go Soft on Buildings Energy Efficiency." September 13.

ENDS Europe. 2005. "Plea for Broader EU Buildings Energy Law." March 16.

ENDS Europe. 2009a. "Council to Clash with MEPs over Efficient Buildings." June 2.

ENDS Europe. 2009b. "EU States Raise Doubts over Greener Building Plans." February 16.

ENDS Europe. 2009c. "States' Fears for Green Building Law Revision Grow." July 3.

ENDS Europe. 2010. "EU Risks Missing Efficiency Goal, Report Warns." September 16.

ENDS Europe. 2012a. "Commission Upbeat about an EED Deal by June," March 8.

ENDS Europe. 2012b. "EED Trialogue Talks Set to Be Very Challenging." March 29.

European Commission. 1979a. "New Lines of Action by the European Community in the Field of Energy Saving." COM 79 (312).

European Commission. 1979b. "Third Report of the Community's Programme for Energy Saving." COM 79 (313).

European Commission. 1984. "Towards a European Policy for the Rational Use of Energy in the Building Sector." COM 84 (614).

European Commission. 1992. "Proposal for a Council Directive to Limit Carbon Dioxide Emissions by Improving Energy Efficiency (SAVE Programme)." COM 92 (182).

European Commission. 1998. "Energy Efficiency in the European Community - Towards a Strategy for the Rational Use of Energy." COM 1998 (246).

European Commission. 2001. "Proposal for a Directive of the European Parliament and of the Council on the Energy Performance of Buildings." COM 2001 (226).

European Commission. 2008a. "Proposal for a Directive of the European Parliament and of the Council on the Energy Performance of Buildings (Recast)." COM 2008 (780).

European Commission. 2008b. "Second Strategic Energy Review: An EU Energy Security and Solidarity Action Plan." COM 2008 (781).

European Commission. 2011. "Proposal for a Directive of the European Parliament and of the Council on Energy Efficiency and Repealing Directives 2004/8/EC and 2006/32/EC." COM 2011 (370).

European Commission. 2015. "Energy Union Package: A Framework Strategy for a Resilient Energy Union with Forward-looking Climate Change Policy." COM 2015 (80).

European Council. 2007. "Conclusions." March.

European Council. 2014. "Conclusions." *EUCO169/14*. October.

Falkner, G. 2016. "The EU's Problem-solving Capacity and Legitimacy in a Crisis Context: A Virtuous or Vicious Circle?" *West European Politics* 39 (5): 953–970. doi:10.1080/01402382.2016.1186386.

Fawcett, T., and G. Killip. 2019. "Re-thinking Energy Efficiency in European Policy: Practitioners' Use Of 'Multiple Benefits' Arguments." *Practitioners' Use of 'Multiple Benefits' Arguments". Journal of Cleaner Production* 210, 1171–1179. doi:10.1016/j.jclepro.2018.11.026.

Henningsen, J. 2011. "Energy Savings and Efficiency." In *Towards a Common European Union Energy Policy: Problems, Progress, and Prospects*, edited by V. L. Birchfield and J. S. Duffield, 131–141. New York: Palgrave MacMillan.

Herranz-Surrallés, A., Solorio, I. and Fairbrass, J. 2020. "Renegotiation Authority in the Energy Union: A Framework for Analysis." Journal of European Integration 42(1), forthcoming..

Italian Chamber. 2011. *Final Document Published Pursuant to Rule 127 of the Rules of Procedure, and Relating To: Proposal for a Directive of the European Parliament and of the Council on Energy Efficiency*. Rome: Italian Chamber.

Keating, D. 2018. "MEPs and Governments at Standoff over Energy Efficiency." EURACTIV, 14 June. Accessed 13 November 2019. https://www.euractiv.com/section/energy/news/meps-and-governments-at-standoff-over-energy-efficiency/

Maltby, T. (2013). "European Union energy policy integration: A case of European Commission policy entrepreneurship and increasing supranationalism". Energy Policy, 55 (April 2013): 435–444. doi:10.1016/j.enpol.2012.12.031

Morgan, S. 2017. "Building Law Gets Stamp of Approval from European Parliament." EURACTIV, 12 October. Accessed 13 November 2019. https://www.euractiv.com/section/energy/news/buildings-law-gets-stamp-of-approval-from-european-parliament/

Oberthür, S., and M. Pallemaerts. 2010. "The EU's Internal and External Climate Policies: An Historical Overview." In *The New Climate Policies of the European Union: Internal Legislation and Climate Diplomacy*, edited by S. Oberthür and M. Pallemaerts, 27–63. Brussels: VUB Press.

Oberthür, S. 2019. "Hard or Soft Governance? The EU's Climate and Energy Policy Framework for 2030." *Politics and Governance* 7 (1): 17–27. doi:10.17645/pag.v7i1.1796.

Oberthür, S., and C. Roche Kelly. 2008. "EU Leadership in International Climate Policy: Achievements and Challenges." *International Spectator* 43 (3): 35–50. doi:10.1080/03932720802280594.

Pollack, M. A. 1994. "Creeping Competence: The Expanding Agenda Of The European Community." *Journal of Public Policy* 14 (2): 95–145. doi:10.1017/S0143814X00007418.

Simon, F. 2018. "'Primary' Energy Efficiency in Focus as EU Talks near Finishing Line." *EURACTIV*, 8 June. Accessed 13 November 2019. https://www.euractiv.com/section/energy/news/primary-energy-efficiency-in-focus-as-eu-talks-near-finishing-line/

Power, authority and security: the EU's Russian gas dilemma

Andreas Goldthau and Nick Sitter

ABSTRACT

This paper investigates contestation of authority in EU energy policy, with a focus on natural gas. It argues that the main challenge centers on the EU's goals and means of energy security policy, not the location and scope of authority. The contested choice is between an across-the-board approach to regulation (Regulatory Power)- and a strategy that opens for the use of regulatory tools for geo-political purposes (Market Power). Competing claims of authority and competing views on how the European Commission should wield its regulatory power reflect both geography (North-Western versus (South-Eastern Europe) and the policy paradigm (market versus geo-politics). The Commission's traditional strategy in energy policy – power-sharing and compromise – only works if there exists a consensus on the ultimate purpose of regulation. However, when the contested issue is whether the Commission should use is regulatory power to pursue market integration or geo-political goals, this presents a genuine policy dilemma.

"In this world, with great power there must also come – – great responsibility"

Spiderman, 1962, Marvel Comics

Introduction

The European Union has considerable power and authority in energy policy. Although the sector features shared competences between the EU and the member states, the EU has limited financial resources, and no control over how member states choose to exploit natural resources, the EU is a prime regulatory force – wielded in no small part by the European Commission. In the 1990s and early 2000s, the Commission directed its regulatory power mainly at building energy markets and improving their workings. The most important policy debates here were about public service provisions and the management of natural resources. However, in the decade following the Russian-Ukrainian gas disputes of 2006 and 2009, the security of supply rose to the top of the policy agenda. The central question was not so much *whether* Vladimir Putin's Russia represented a potential enemy armed with an 'energy weapon', as to *how* the EU could and should use its regulatory power in pursuit of the goal of ensuring affordable and reliable external gas supplies. Consequently, the core issue is not so much whether the EU *has* the power and authority to use regulatory policy tools, as the *purpose* for which the EU should wield its power.

The central theme that unites the articles in this special issue is how authority is contested in EU energy policy. As the introductory article (Surrallés, Solorio, and Fairbrass 2020) makes clear, authority is a somewhat broader concept than either power or competence. The global governance literature has demonstrated that author-ity – the ability to make decisions that others accept as legitimate and comply with – is in constant flux. It is open to contestation; there can be overlaps between competing claims to authority; and authority can be claimed by a variety of actors. In the EU context, this means that even when legal competences are established, authority can be contested. Nevertheless, because the EU political system is ultimately dependent on the member states for enforcement and policy implementation, the three concepts of *power, authority* and *competence* are closely linked. If the European Commission's authority is contested, this raises questions about its competence, which in turn are likely to affect its power. Conversely, in the effort to build a single energy market over the last three decades, the Commission increased its competence and authority in parallel. On several occasions, it explicitly refrained from threatening to use its legal power (under competition law) to go beyond what it saw as limits to member state consensus (which conferred authority on the Commission).

Since the Commission embarked on its quest to build a single European energy market in the early 1990s, three main issues have caused debates about the appropriate limits of EU competence, and have been the source of overlapping claims of authority in the energy field. The first is the appropriate balance between member state and European Union competences. The second is the balance between the priorities of competition policy, environmental policy, and security of supply. Both are addressed in other articles in this volume. The focus here is on the third source of contested authority: the strategic dimension of energy policy. Energy is not simply a good like any other. It is a mixed good in that it has some of the characteristics of a private good (it is rival and excludable in consumption) and some of a public good. The stable supply of energy at socially acceptable prices is one aspect of the public goods dimension, and it has strategic implications.

Although the security of supply question is pertinent to all sources of energy, it is most salient with respect to natural gas. Whereas oil markets are global, a substantial part of the natural gas trade in Europe is still conducted in terms of bilateral contracts. Russia is the dominant supplier for many states, and the appropriate role of 'red gas' (Högselius 2012) is very much contested. With the Russian annexation of the Crimea in 2014, the external dimension of EU energy policy became even more salient, and security of gas supply emerged as a core issue in the strategy for an Energy Union adopted by the Commission in February 2015. It has been argued that the step towards the Energy Union signaled a shift from a priority of market liberalization in the three regulatory 'energy packages' of 1998, 2003 and 2009, to security of supply (Andersen, Goldthau, and Sitter 2017b), in addition to tying EU energy policy more firmly back to climate goals. When the Commission proposed a revision of the Gas Directive in November 2017, to 'clarify' that the core principles of EU energy legislation apply to gas pipelines to and from third countries up to the EU border (European Commission 2017a), this was in a context where some member states were at odds over whether the planned Nord Stream 2 pipeline from Russia to Germany constituted an energy security risk (Loskot-Strachota 2016), and whether it could or should be blocked. This debate goes to the heart of the problem of EU authority with respect to external gas trade. It is not so much about where power or competence should be located, as much as the goals for which power should be exercised. In its simplest form, the question is whether the EU ought to use its regulatory power to address a threat that arises from geopolitics.

The present article explores how the EU Commission has dealt with this question. To this end, we juxtapose the concepts of Regulatory and Market Power in order to delineate a market-liberal from a geo-political approach to regulation. This allows tracing the Commission's deliberate move from focusing on advancing gas market integration to becoming an actor in the realm of EU external security policy by at the same time unveiling the underlying contestations. With this, the present article zooms in on what (Surrallés, Solorio, and Fairbrass 2020) call conflicts of authority on the horizontal dimension, which arise from a recalibration in the principal paradigm underpinning EU policy-making in natural gas. At its core, the EU's regulatory power rests on the market-liberal model. Using regulation to address issues other than those concerned with the design and functioning of commodity or other markets – such as geopolitics – not only crosses the line to market power. It also inevitably creates tensions. As the present article argues, the Commission as the contending actor adopted a legal strategy to sanction its advance into new policy territory. On one hand, this strategy addresses sovereignty-based contestation, as it amends the Commission's authority. On the other, it also addresses substance-based contestation, as it positively sanctions a geopolitical interpretation of the EU regulatory toolbox.

This article is divided into four parts. The first assesses the Commission's competences and authority with respect to the EU's external gas trade. It shows that the gas market liberalization which got underway in the late 1990s has resulted in a policy regime with some overlapping competences, which the Commission to a large extent managed through power-sharing. The second section turns to how authority has been contested in the gas sector, using the Nord Stream 2 debate as a case. Here, the Commission has considerable power, but its authority is dependent on its legitimate – or responsible (in the eyes of the beholders) – exercise of this power. The third part turns to the Commission's options in the face of open contestation between liberally and geopolitically oriented member states. It shows that when it comes to dealing with the security of external gas supply, the choice between market-oriented and strategic use of regulation is a genuine policy dilemma. The fourth section turns to analyzing how the Commission has used regulation to try to avoid or circumvent this dilemma, and tried to strike a balance between a regulatory power strategy and a market power strategy. Finally, the conclusion reflects on the theoretical implications of the findings. It also puts forward the more normative argument that in the choice between regulatory power and market power, only the former is compatible with the EU's long-standing grand strategy.

Power, authority and consensual decision-making in EU gas market policy

Like many other EU policy sectors, the allocation of power, authority, and policy-making in the energy sector is the outcome of a gradual process that has involved both hard bargains between the member states and a gradual extension of the Commission's formal competence. What makes energy policy – and natural gas policy in particular – somewhat exceptional is that it also involves compromises between three very different policy goals: free trade, sustainability, and security of supply. The EU member states have long been divided as to the nature and importance of the three issues, and these divisions sometimes cut across each other. For example, both the UK and the Nordic states have been more market-oriented since the early 1990s, but with the latter also taking a more assertive stance on the environment dimensions of energy policy (Liefferink and Andersen 1998). Likewise, both Poland and Hungary are somewhat skeptical toward the free-market dimension of EU energy policy, but they take radically opposite views on whether reliance on Russian gas constitutes a security problem or not (Nosko and Mišík 2017).

When it came to market liberalization, the Commission's strategy in the 1990s and 2000s was shaped by its reluctance to use its formal powers in the face of member state opposition. The Commission had considerable formal powers (i) in Robert Dahl's sense of compelling other actors to do something (Dahl 1957); (ii) in Schattschneider (1960) and Bachrach and Baratz (1962) sense of setting agendas and establishing what the alternatives are; and even (iii) in Lukes' (1974) sense of influencing what other actors want. But the Commission's authority – in the Weberian sense of the commands of one actor being accepted as binding by the recipient (Spencer 1970) – depended very much on its actions being seen as legitimate in what Weber (1922) called a legal-rational sense. Indeed, the Commission's quest to maintain and enhance its long-term authority in the energy sector caused it to promote compromise arrangements for the first energy packages that were the subject of open debate between the energy and competition Commissioners (Andersen and Sitter 2006). Consequently, the Commission won acceptance for the principle of market liberalization at the cost of exemptions and derogations that allowed the more sceptical member states to claim that they kept liberalization at bay in practice. Over the next decade, the Commission persuaded the member states to adopt follow-up legislative packages that pushed market liberalization in the gas sector forward, at the same time as the boundary between national and EU competence was elaborated more clearly in the 2007 Lisbon Treaty.

The Lisbon Treaty reflects the fact that national governments have traditionally regarded the management of natural resources and policy decisions about the use of energy resources as strategic for national welfare and/or economic development. These preferences were hardwired into the Treaty by way of Art. 194 (2), which states that: '[EU level energy] measures shall not affect a Member State's right to determine the conditions for exploiting its energy resources, its choice between different energy sources and the general structure of its energy supply, without prejudice to Article 192(2)(c)' (European Communities 2007). Indeed, Art. 192(2) states that environment policy decisions shall be subject to unanimity requirements in the Council of Ministers in the case of 'measures significantly affecting a Member State's choice between different energy sources and the general structure of its energy supply.' Even though directives on renewable energy have been adopted under majority voting procedures, Articles 192 and 194 thus explicitly protect the member states' rights to choose their own energy mix.[1]

On the other hand, as gas markets were liberalized after 1998, energy firms and utilities were subject to ordinary EU competition law as well as the specific rules laid down in the three energy packages of 1998, 2003 and 2009 (European Parliament and the Council 1998, 2003, 2009). Combined, these directives established third party access (TPA) to pipelines, legal and ownership unbundling of distributing and transmissions activities, and set up the independent Agency for the Cooperation of Energy Regulators (ACER). From the Commission's perspective, this was a triumph of the idea of a liberal internal market, even if the road there was not completely smooth. Conceptually, internal EU energy policy, therefore, mirrors, above all, the regulatory state model for governance (Lodge 2008; Majone 1994). It is concerned primarily with 'creating markets and making them work' (Goldthau and Sitter 2014, 2015a). Even though environmental policy emerged as a very important policy area and directly affects energy markets, the main principle remains liberal. Likewise, the Security of Supply Regulation (European Parliament and the Council 2010) was primarily a directive that supplemented the market-based approach, and hardly signaled a change of policy or direction. The 2015 Energy Union strengthened the focus on sustainable development and security of supply (as well as research, innovation and competitiveness), but it does not change either the formal competences of the EU and its member states or the practice (Andersen, Goldthau, and Sitter 2017a; Szulecki et al. 2016).

The empirical question is whether these patterns of mixed authority caused serious problems in terms of contested authority, sovereignty or efficient policy-making. Until recently, the overall answer was negative. From the early years of the European Economic Community until the mid-1990s, the member states effectively kept control of their energy sectors largely by keeping areas such as gas and electricity outside the scope of EU competences and resisted efforts by the European Commission to change this. Since 1998, the member states have maintained considerable influence of their gas sector through national regulation and ownership, and by ensuring that EU-level decision-making is consensual. As gas market liberalization proceeded, the European Commission and the member states elaborated new mechanisms for maintaining consensus and sharing power. Key mechanisms for managing member state hetero-geneity in the gas sector include temporary derogations, delegation of power to independent agencies (which operate consensually), a degree of national control over implementation (including state ownership of energy companies), and even the Commission adopting a pragmatic stance towards accommodating member state preferences (Andersen and Sitter 2014).

This kind of power-sharing is sometimes discussed as a consequence of or feature of liberal intergovernmentalism (Moravcsik 1993) or new intergovernmentalism (Puetter 2014), but conceptually it owes more to the debates about power-sharing in plural democracies (Lijphart 1977). Consensual decision-making was essential to produce out-comes that the states could accept as legitimate and was thus needed to establish the Commission's *authority* in the sector. Two decades later this logic still held: the delegation of power to ACER involved a Board of Regulators consisting of senior representatives of the member states' National Regulatory Authorities. As a consequence, and although both new legislative proposals and individual regulatory decisions keep on being contested in terms of content, there has been little contestation in terms of rival claims to sovereignty or authority.

Contesting EU authority in the gas sector

To be sure, EU gas market integration was not a linear process. For a variety of reasons, many member states failed to fully implement the various energy packages. But as the Commission's 2014 review of the EU's 'Progress towards completing the Internal Energy Market' (European Commission 2014) showed, most of these problems owed to difficulties with coordination and the need to elaborate new technical standards. As in any other EU sector, there is a regular supply of infringement cases, but few involve any kind of dispute about the scope of the Commission's authority. Similarly, the Commission's 2016 report on the compliance of bilateral Intergovernmental Agreements (IGAs) between EU member states and non-EU energy suppliers (European Commission 2016) demonstrated a clear lack of member state action. However, the problem was not related to the legitimacy of the Commission evaluating IGAs which is what the Commission had been asked to do as per a 2012 Decision in order to inject more transparency into the market (European Parliament and Council of the European Union 2012); but rather to legal difficulties arising from the ex-post nature of the compliance check.

The main challenge to power-sharing as a mechanism for securing legitimacy and enhancing authority has come with the increased salience of *energy security*. This is not so much of a question of where formal and informal power lies, as of the goal for which power is exercised in the first place. This issue is rooted in a deeper division between states that take a liberal approach to gas markets and those that see gas primarily through a geopolitics lens. Since the turn of the millennium, a political cleavage has developed between the liberal EU and European Economic Area states that view gas mainly as a private good with some of the ordinary public goods characteristics of a network industry, and those that see gas as an important strategic good and fear that Russia might deploy the 'energy weapon'. For the former, energy remains a matter of low politics though the security argument may occasionally help pushing the cause of market integration (Herranz-Surrallés 2015). The latter view it as a question related to high politics (Andersen, Goldthau, and Sitter 2017b) and securitization (Boersma and Goldthau 2017; Judge and Maltby 2017; Natorski and Herranz-Surrallés 2008). Judging by their official public position on core energy policy priorities, the most prominent and vocal members of the first group include Germany, the Netherlands, the UK, the Czech Republic and Norway; the second is led by Poland and the three Baltic states. The disputes over the desirability of the North Stream 2 gas pipeline from Russia to Germany through the Baltic Sea and how best to deal with this issue by and large reflected this cleavage (see Table 1).

The issue at hand has three dimensions, which add up to a cleavage between liberal and geopolitically oriented states. The first dimension concerns the nature of the Russian threat to the EU (and EEA), and the role energy plays in this. The main difference between the two groups of states is that the more liberally oriented acknowledge the Russian threat, but do not see this as necessarily having a major gas dimension. The second dimension concerns what should be the main concern for EU energy policy: security or growth. Here the liberal states separate energy from security policy, while the more geopolitically oriented ones see these two issues as inseparable. Consequently, the

Table 1. Official public stakeholder positions on the contested Nord Stream 2 gas pipeline, 2017/18.

Country/Actor	Role (or presence of firm)	Position in 2017/2018	Use EU regulation to prevent NS2?
Russia	Terminal	Strongly favoured NS2	No
Germany	Terminal	Favoured NS2, but increasingly concerned about geopolitical impact; governing parties internally divided	No
Finland	Transit EEZ	Against NS2 for geopolitical reasons, but resultantly approved transit	No
Sweden	Transit EEZ	Against NS2 for geopolitical reasons, but resultantly approved transit	No
Denmark	Transit EEZ and territorial waters	Against NS2 for geopolitical reasons, but resultantly approved transit, passed new law to permit blocking NS2 on security grounds	No
Poland	Espoo consultation	Strongly opposed to NS2 for geopolitical reasons, blocked consortium in 2016	Yes
Latvia	Espoo consultation	Opposed to NS2 for geopolitical reasons	Yes
Lithuania	Espoo consultation	Opposed to NS2 for geopolitical reasons	Yes
Estonia	Espoo consultation	Opposed to NS2 for geopolitical reasons	Yes
Norway	Kvaerner (subcontractor)	Took no position, because the country is Russian main competitor in EU gas supply	No
France	Engie	Moderately favourable, focus on trade	No
Netherlands	Shell	Moderately favourable, focus on trade, but highly critical of Russian Ukraine policy	No
UK	Shell	Against NS2 for geopolitical reasons	Yes
Austria	OMV	Moderately favourable, focus on trade	No
Hungary	Energy deals with Russia	Pro-Russian, but solidarity with Poland, ambivalent and low-profile on NS2	No
Czech Rep		Somewhat pro-Russian, ambivalent and low-profile on NS2	No
Slovakia		Against NS2 for geopolitical reasons	Yes
Bulgaria		Somewhat pro-Russian	No
Italy	TAP terminal	New government opposed to TAP, and friendly to Russia	No
EU Energy Commissioner		Against NS2 for geopolitical reasons and security of supply reasons	Yes
EU Council President		Against NS2 for geopolitical reasons and security of supply reasons	Yes
Other EU states		More or less neutral, East Central European states more sympathetic with Poland	Divided
USA Congress		Strongly opposed to NS2 for geopolitical reasons	N/A
USA President		Strongly opposed to NS2 as potential competitor in gas supply (LNG)	N/A
NATO General Secretary		No position, NS2 is not a NATO issue	N/A

Source: Comprehensive survey of DW, EUobserver, Euractiv, Financial Times, Norwegian Telegram Bureau, Oil & Gas Journal, Politico, Politiken, TASS, for 2017–2018 (August); compilation authors' own.

third dimension of the energy security issue becomes contestation over how the EU (and its member states) should use the regulatory tools at their disposal. The next section turns to the strategies available to the EU for dealing with this question of how and for what purpose its regulatory authority should be used in the gas sector.

Managing contested authority in the gas sector: the European Commission's policy options

In the gas sector, the EU has the tools to manage diffuse authority as far as the internal dimension of energy policy is concerned. It does so by power-sharing between the member states and the EU level, using various ways of delegating power to specialized agencies, embracing modes of private regulation (see Eckert and Eberlein 2020), or even experimentalist governance (Eberlein 2010). During the 1990s and early 2000s, in the processes leading up to the first and second energy packages, a certain degree of 'fuzziness' (Andersen and Sitter 2009) about the goals was both possible and desirable. Both the enthusiastic liberalizers (e.g. the UK and the Netherlands) and the more skeptical ones (e.g. France and Belgium) could present the compromise solutions and gradualist approach to liberalization as a political victory. Compromise might limit the extent of reform at any given date but left the road open for more reform later. In the 1990s the Commission's principled push for energy liberalization (driven by a series of its Competition Commissioners) was tempered by a pragmatic approach to gas markets (advocated by Energy Commissioners) designed to avoid abrupt change. In the 2000s, this pragmatic liberalization continued. Power-sharing is not much good until everyone agrees what this power is used for, but in the case of the extension, the Single European Market to the gas sector the Commission managed to secure agreement on the overall goals in return for compromise on the practicalities of liberalization. But when it comes to the external dimension of energy policy, there is no such consensus.

The starting point for a discussion of the Commission's options for dealing with contestation over the security dimension of its gas policy is that (unlike gas liberalization) this is no longer the kind of problem related to competing sovereignty claims or diffuse authority that can be prevented by managing the transition from less to more EU authority. As long as Russia was seen as a reliable energy supplier, the security dimension of the EU's gas trade with its big neighbour had low salience. However, this began to change with the Russian-Ukrainian gas disputes of 2006 and 2009, both of which affected gas supply in south-eastern EU member states. In 2014 the Crimea crisis increased the salience of this issue dramatically, and the downing of Flight MH17 over Ukraine on 17 July 2014, made the question of how the EU should react acute (193 of the 298 casualties were Dutch nationals). Reports of Russian interference in the US elections and the Brexit referendum of 2016 hardly reduced concerns about a resurgent Russian imperialism. In addition, both the EU and NATO suffered internal stress, the former linked to the populist challenges to the rule of law in countries like Hungary and Poland, and the latter when the new US president questioned his commitment to collective defense. In this deteriorating geopolitical context, gas trade was just one of many dimensions of EU security that rose to the top of the political agenda. It could no longer be managed by downplaying or fudging the issue.

If the EU's problems related to competing claims about how authority ought to be used in the gas sector cannot be prevented, can they be managed? Conceptually, there is a fundamental difference between management by compromise and power-sharing, on one hand, and management by way of stark choices on the other. In the gas market liberalization process, power-sharing went hand in hand with delays, compromise and de-politicization. But when it comes to the security dimension of gas, power-sharing and de-politicization does do not offer viable routes out because it is the politicization of gas trade that is the root of the problem. The challenge for the Commission lies in how to *wield* power. This gets us to an important conceptual aspect of the EU energy security conundrum: how the different claims of authority loop back to the way EU authorities use the power at their disposal, the purpose of wielding this power, and the paradigm underpinning it. In policy terms, this is a straightforward dilemma: the Commission can either use its regulatory authority for a wider set of political ends, or it can limit it rather narrowly to making the EU gas market work better. As far as the external dimension of gas regulation is concerned, managing problems of authority means choosing one strategy or another.

As discussed in more detail elsewhere (Goldthau and Sitter 2018), there are four ways the EU can wield external power in energy affairs: (1) classic *neutral power of regulation* (to build markets and manage them), (2) *regulatory power* entailing a bias toward consumer (shifting the balance between exporters and importers, for EU consumer benefit), (3) *market power* (featuring a selective use of regulation, notably for political purposes), and (4) hard *economic power* (making energy – which in many ways can be conceptualized as low politics – subject to high politics rationales). All four types of wielding power lie on a continuum from a liberal approach to a geopolitical approach (see Figure 1).

At the EU-level, strategies based on the models on both ends of the spectrum have proven difficult. The fully liberal approach relies on a broad agreement on the desirability of international rules-based governance, which is in short supply beyond the EU's borders. The hard economic power model, on the other hand, is applicable in very limited circumstances only (such as the Iran sanctions), as it rests on a common foreign and security policy and the collective action of EU member states. By contrast, the EU has

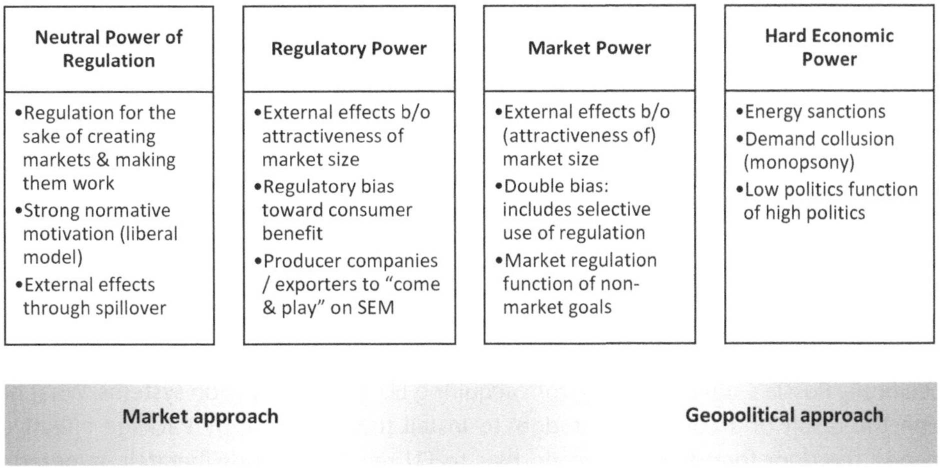

Neutral Power of Regulation	Regulatory Power	Market Power	Hard Economic Power
• Regulation for the sake of creating markets & making them work • Strong normative motivation (liberal model) • External effects through spillover	• External effects b/o attractiveness of market size • Regulatory bias toward consumer benefit • Producer companies / exporters to "come & play" on SEM	• External effects b/o (attractiveness of) market size • Double bias: includes selective use of regulation • Market regulation function of non-market goals	• Energy sanctions • Demand collusion (monopsony) • Low politics function of high politics

Market approach   **Geopolitical approach**

Figure 1. EU types of external power in energy.

proven very successful when it comes to building an internal energy market and regulating its functioning. Given its size, it has also become evident that there exists a strategic and indeed external dimension of EU gas market regulation. Unsurprisingly, therefore, it is around the deep contestations about the nature of the Russian threat, what the EU is ultimately for (security or growth), and how the Commission should use its (limited but elaborate) regulatory state toolbox that the two models of regulatory versus market power have clashed. This is what we turn to next.

Confronting the security of supply policy dilemma: regulatory power or market power?

EU member states that subscribe to a liberal model and a pro-growth agenda tend to advocate the *regulatory power* approach when it comes to addressing security of supply concerns This is notwithstanding their stance toward Russia. As Table 1 suggests, member states that fall into the 'liberal' camp tend to have a strong view on the Kremlin's geopolitics, including the Netherlands or Sweden. Still, their preferred approach to dealing with Gazprom, the state-owned gas monopolist, and Russian gas more generally, is one based on Single Market rules. From a regulatory power perspective, Gazprom becomes subject to the entire body of EU energy regulation the moment the company's energy services enter the SEM. Gazprom will be forced to comply with the EU's (unilaterally set) rules, as any other company serving EU customers. In other words, the way the Russian gas exporter would be treated is as a dominant market player – not as Moscow's foreign policy arm.

In line with the regulatory power model, the Commission – the EU's chief market authority – enforced the end of destination clauses in gas contracts, battled against long-term take-or-pay agreements and launched a competition policy case against Gazprom for abuse of its dominant position in gas pricing (Goldthau and Sitter 2015a). Arguably, the Commission acted in an increasingly assertive manner, determined to use its powers to the fullest. The decisive element here, however, lay in the fact that the Commission would target *any* company that sold gas to the EU, whether from the EEA or Russia, and in the case of long-term contract LNG even from Nigeria (Talus 2011). Put differently, when exercising regulatory power, even when 'playing tough', the EU still by and large applied energy regulation in a neutral way, across the board toward all (external) actors it deemed to be possibly violating EU rules. The sole regulatory bias consisted in favoring (domestic) consumer interests over (foreign) producer interests. It was not so much the 'nationality' of each foreign suppliers that made the difference, as their character and behavior.

Arguably, the Commission acted in line with the regulatory power model because it is a 'liberal animal' by design (Goldthau and Sitter 2015a). But it also did so because some other, arguably more strategic, elements that had found their way into EU regulation ensured the continued support of more geopolitically minded member states. For instance, the 2009 Third Energy Package introduced the much-debated 'Gazprom clause' (Cottier, Matteotti-Berkutova, and Nartova 2010) – a thinly veiled attempt to prevent specifically Russia's state company from acquiring EU gas transmission systems. What had been the result of a consensual 'fudge' to install the Third Energy Package effectively opened the door for adding a second bias to EU regulation: a deliberately targeted or selective use thereof, crossing the line to a market power approach.

The most prominent case in this regard is the Nord Stream 2 pipeline, the 55 bcm offshore link between Russia and Germany across the Baltic Sea. Nord Stream 2, owned and built by Gazprom, will double the capacity of the existing Nord Stream link, thus possibly cementing Russia's dominant position in the European gas market, and put in question future transit of Russian gas through Ukraine. The project is politically contested, not the least against the backdrop of Russian gas supplies being subject of long-standing East European fears of import dependency, but it also owes much to Russia's annexation of Crimea in 2014 (for a detailed discussion, see Schmidt-Felzmann 2020). The strategy of the Commission, an early critic of additional pipelines bringing Russian gas into the EU, has been to make Nord Stream 2 subject to the Third Energy Package and therefore to TPA provisions, which the Gazprom-led infrastructure project would find hard to comply with.

This process saw several twists and turns. As applying existing TEP regulations to offshore pipelines proved legally not waterproof, the Commission in late 2017 suggested amending the 2009 Gas Directive and extending it to gas pipelines to and from third countries (European Commission 2017b). The proposed revision envisaged derogations for existing pipelines and exemptions for planned projects. A flanking Memo issued by the Commission, however, detailed that among the two planned upstream pipelines affected by the proposed revision, only Nord Stream 2 would need to fully comply, as the Trans-Adriatic Pipeline (TAP) connecting Greece's Turkish border with Italy was granted an exemption (European Commission 2017c). After a year-long tug-of-war between proponents and opponents of the Commission proposal, the Council and the European Parliament in February 2019 agreed on a compromise de facto extending EU TEP provisions to offshore pipelines, but ensuring legal oversight remained with the Member State in whose territorial sea 'the first interconnection point with the Member States' network is located' (Council 2019). The Commission will retain the role of vetting the legal decisions of that very member state, including ruling on exemptions from TEP rules that said member state (Germany in the case of Nord Stream 2) might ask for. The Gazprom-owned Nord Stream 2 company responded by exploring options for corporate restructuring to circumvent or limit Commission oversight, including establishing a separate company to manage the part of the pipeline that would run through German territorial waters (FT 2019). The Gas Directive Amendment was adopted on 15 April 2019.

It is fair to argue that the rationale underpinning the legal debate in the EU was to stop the project on grounds of not being compatible with EU energy laws, rather than to ensure its lawful operation. This entails a distinct element of selective, rather than equal, treatment, in which regulation as applied singles out individual targets rather than applying across the board. The stated objective of such a market power approach is to prevent the project in favor of a politically preferred supply route – across Ukraine – and to strengthen the Strategic Partnership on energy with Kyiv (European Commission 2015). More generally, it reflects a broader, geopolitical motivation, in which Russian-sponsored pipeline infrastructure becomes a mere proxy for reacting to Moscow's assertive foreign policy. With this, EU energy regulation is meant to serve national security goals, not necessarily market-related ones. This contrasts with a liberal, regulatory power approach that would seek to maintain competition in the EU downstream gas sector and strengthen market oversight in light of the pipeline's potential impact on market liquidity levels or price structures (for a critical discussion of this aspect, see Kotek, Selei, and Tóth 2017).

The proponents of a market power approach toward European pipeline politics, and particularly to Nord Stream 2, are indeed primarily found in Eastern Europe, thus confirming the split between more market-oriented, liberal EU member states, and the ones concerned primarily with national security as regards their external gas supplies (Lang and Westphal 2017). But it is important to note that the rift runs all the way through Europe. A case in point is German politicians arguing in favor and against Nord Stream 2 across the political spectrum, and resting on opposite views of Nord Stream 2 being an economic or a political project (Caspary et al. 2018; Weber et al. 2018). In other words, competing claims of authority, and competing views on how EU authorities should wield their power, may to a certain degree reflect both geography and the paradigm typically represented by it (North-Western Europe representing a more market-focused, liberal outlook, Eastern and South-Eastern Europe a more security-focused one). At the same time, the divisive nature of Russia's foreign policy action, particularly in Ukraine, gives rise to deep contestations over the appropriateness of sticking to a regulatory power model, in the face of mounting geopolitical pressure on Europe – contestations which do not necessarily follow geography.

An interesting case in this regard is Denmark. The Danish parliamentary debates about how to deal with Nord Steam 2 resulted in a new law that obliged the foreign minister to consider the security implication of the pipeline project (if it ran through the country's territorial waters), while the environment impact assessment was left to the Danish Energy Agency (Folketingstidende 2017). The Nord Stream 2 company responded by exploring alternative routes that ran through the Danish Exclusive Economic Zone but avoided its territorial waters. It submitted a proposal for this in August 2018, over which the Danes procrastinated in the hope that a decision might be taken at the EU-level. With the February 2019 EU compromise and in the light of the new EEZ-only route, the Danish territorial sea issue became less important. On 30 October 2019, the Danish Energy ministry approved the new route as an 'administrative decision' – at considerable costs to the project in terms of both time and money. The delay also meant that the pipeline would be covered by the new above-mentioned EU Gas Directive. The important point here is that Denmark separated security from regulatory aspects, thus eschewing the Market versus Regulatory Power debate.

Conclusion

Great power brings great responsibility. In energy security, the main challenge for the EU in terms of competing claims to sovereignty and authority is about the goals and means of energy policy. For the Commission, authority means legitimacy, and this depends on the responsible use of power. As the discussion above suggests, the contestations surrounding external gas policy do not so much center on the Commission's legal competence, as on what 'appropriate use of power' means, i.e. what ends regulation should be used for. This contestation culminated in the debate around Nord Stream 2.

The Nord Stream case 2 speaks to two central elements of the horizontal contestation outlined by (Surrallés, Solorio, and Fairbrass 2020) the sovereignty element, i.e. the formal authority of the Commission to act in the case of an offshore pipeline; and the substance-related element, which essentially is about redefining the purpose of EU regulation inspired by geoeconomics rather than being an exercise of market-liberalism. The interesting finding here is that the Commission obviously used a legal sanctioning strategy for both, to do away with ambiguity in authority and to sanction its foray into external security affairs. This warrants further empirical investigation into the way the legal strategy may be used for substance-related contestations of authority in the European Union, using cases from policy fields other than energy. Moreover, the findings call for further inquiry into the degree to which legal sanctioning fosters (or not) the de-politicization of contested issue areas. The Commission's strategy to have EU member states formally vet its (new) powers in external energy infrastructure regulation arguably represented a deliberate step to further politicize Brussel's prime toolbox, not the opposite. Additional research is therefore needed to explore whether the de-politicization argument holds only in certain contexts, notably under the conditions of the liberal paradigm.

In addition to the theoretical implications, the discussion in this article raises normative questions and highlights trade-offs entailed in the EU's current trajectory. Overall, three conclusions can be drawn. First, as gas trade is becoming highly (geo-)politicized, this puts an end to fuzziness as the workhorse of EU gas politics. The deep division over how to cope with Russian gas cannot be managed by power-sharing and a measured accommodating of member state preferences. The Commission faces a real policy dilemma: it will be called upon again and again to use regulation to keep in check projects that some governments see as a geopolitical problem. Accommodating these calls would change the nature of the EU's regulatory regime – at least in terms of how regulation is used in practice. In the context of Nord Stream 2, the Commission faces the choice between a regulatory power or a market power strategy. The 2019 amendment opens the door to the Commission openly using regulation in an attempt to halt the pipeline, or linking (the threat of) regulatory action to its effort to broker a Russia–Ukraine gas deal. Such an exercise of market power would politicize regulation, and test the limits of the EU's 'soft power with a hard edge' (Goldthau and Sitter 2015b).

Second, there is a risk that a market power approach to gas regulation undermines the central purpose of regulation – to build and manage a single market for energy and a level playing field, broadly acceptable to all member states. The contestation around the main purpose of EU energy regulation – market creation or geo-economics – goes straight to the heart of a very normative question: the nature of EU as a polity. The EU, like all states, has a grand strategy: it is a liberal actor. Regulatory power, both as a normative concept and as it is practiced, is compatible with this grand strategy; market power, not so much. Moving toward a market power approach might, therefore, challenge the EU's grand strategy – which is an important source of its legitimate authority. Ultimately, this involves questions of legitimacy. Member states that favor contested Russian pipeline projects view the role of the EU as limited to ensuring market functioning, and tend to question the legitimacy of EU authorities' interfering in these projects for security reasons. States that oppose these pipelines view the role of EU authorities as going beyond market aspects, and EU energy regulation as a perfectly legitimate means to address geopolitical challenges. Leaving the regulatory power/market power dilemma unresolved might decrease the EU's authority in the eyes of both types of states.

Third, and as a consequence, the EU might be well advised to address security problems by means of security policy tools, rather than dealing with such challenges through regulation. This means 'properly' securitizing such issues, with the consequence of them becoming subject to the hard security toolbox.[2] As discussed above, in fact, the Danish government showed a possible way out of the gas security/trade conundrum: instead of using environment rules for preventing the project, they passed a new law allowing for assessment of pipeline projects that cross their national waters on national security grounds. If a similar approach were taken at the EU level, this might come at the cost of the EU not always being able to act. But it would solve a problematic policy dilemma by clearly delineating what currently gets muddled: regulatory power for regulatory purposes; and hard economic power for security purposes. The danger of crossing the regulatory power/market power is that 'irresponsible' use of power might undermine the EU's authority in the long term.

Notes

1. For the potential of shale gas as a domestically available energy resource particularly in Eastern Europe, see Goldthau (2018).
2. For a securitization lens on EU energy policy, see Szulecki (2017).

Disclosure statement

No potential conflict of interest was reported by the authors.

ORCID

Andreas Goldthau http://orcid.org/0000-0001-9814-6152

References

Andersen, S., A. Goldthau, and N. Sitter. 2017b. "From Low to High Politics? the EU's Regulatory and Economic Power." In *Energy Union. Europe's New Liberal Mercantilism?* edited by S. Andersen, A. Goldthau, and N. Sitter, 13–26. Basingstoke: Palgrave Macmillan.

Andersen, S., A. Goldthau, and N. Sitter, eds. 2017a. *Energy Union: Europe's New Liberal Mercantilism?* Basingstoke: Palgrave Macmillan.

Andersen, S. S., and N. Sitter. 2009. "The European Union Gas Market: Differentiated Integration and Fuzzy Liberalization." In *Political Economy of Energy in Europe*, edited by G. Fermann, 63–84. Berlin: BWW.

Andersen, S. S., and N. Sitter. 2006. "Differentiated Integration: What Is It and How Much Can the EU Accommodate?" *Journal of European Integration* 28 (4): 313–330. doi:10.1080/07036330600853919.

Andersen, S. S., and N. Sitter. 2014. "Managing Heterogeneity in the EU: Using Gas Market Liberalisation to Explore the Changing Mechanisms of Intergovernmental Governance." *Journal of European Integration*. doi:10.1080/07036337.2014.953947.

Bachrach, P., and M. S. Baratz. 1962. "Two Faces of Power." *American Political Science Review* 56 (4): 947–952. doi:10.2307/1952796.

Boersma, T., and A. Goldthau. 2017. "Wither the EU's Market Making Project in Energy: From Liberalization to Securitization?" In *Energy Union. Europe's New Liberal Mercantilism?* edited by S. Andersen, A. Goldthau, and N. Sitter, 99–114. Basingstoke: Palgrave Macmillan.

Caspary, D., W. Langen, J. Pfeiffer, G. Nüßlein, H. Heil, B. Westphal, A. Post, and T. Gremmels. 2018. "Nord Stream 2 Stärkt Europa." *Frankfurter Allgemeine Zeitung*, March 07.

Cottier, T., S. Matteotti-Berkutova, and O. Nartova. 2010. "Third Country Relations in Eu Unbundling of Natural Gas Markets: The "Gazprom Clause" of Directive 2009/73 Ec and Wto Law." In *NCCR Trade Regulation Working Paper*. Bern: World Trade Institute.

Council of the European Union and European Parliament. 2019. "Directive (EU) 2019/692 of the European Parliament and of the Council of 17 April 2019 Amending Directive 2009/73/EC Concerning Common Rules for the Internal Market in Natural Gas (Text with EEA Relevance.)„. *Official Journal of the European Union* L 117/1 (2019).

Dahl, R. A. 1957. "The Concept of Power." *Behavioral Science* 2 (3): 201–215. doi:10.1002/bs.3830020303.

Eberlein, B. 2010. "Experimentalist Governance in the European Energy Sector." In *Experimentalist Governance in the European. Towards a New Architecture*, edited by C. F. Sabel and J. Zeitlin, 61–79. Oxford: Oxford University Press.

Eckert, S., and B. Eberlein. 2020. "Private Authority in Tackling Cross-border Issues: The Hidden Path Of Integrating European Energy Markets." *Journal of European Integration* 42 (01).

European Commission. 2014. *Progress Towards Completing the Internal Energy Market*. COM(2014) 634 final. Brussels.

European Commission. 2015. *Energy Union Package. A Framework Strategy for A Resilient Energy Union with A Forward-Looking Climate Change Policy*. COM(2015) 80 final. Brussels: Communication from the Commission to the European Parliament, the Council, the European Economic and Social Committee, the Committee of the Regions and the EUropean Investment Bank.

European Commission. 2016. *Report from the Commission to the European Parliament, the Council and the European Economic and Social Committee on the Application of the Decision 994/2012/EU Establishing an Information Exchange Mechanism on Intergovernmental Agreements between Member States and Third Countries in the Field of Energy*. COM(2016) 54 final. Brussels.

European Commission. 2017a. *Energy Union: Commission Takes Steps to Extend Common EU Gas Rules to Import Pipelines*. IP/17/4401, 08.11. Brussels.

European Commission. 2017b. *Proposal for a Directive of the European Parliament and of the Council Amending Directive 2009/73/EC Concerning Common Rules for the Internal Market in Natural Gas (Text with EEA Relevance)*. COM(2017) 660 final. Brussels.

European Commission. 2017c. *Questions and Answers on the Commission Proposal to Amend the Gas Directive (2009/73/EC)*. MEMO 17/4422. Brussels.

European Communities. 2007. "The Treaty of LisbonAmending the Treaty Establishing the European Union and the Treaty Establishing the European Community Including the Protocols and Annexes, and Final Act with Declarations." Lisbon.

European Parliament and Council of the European Union. 2012. "Decision No 994/2012/EU of the European Parliament and of the Council of 25 October 2012 Establishing an Information Exchange Mechanism with Regard to Intergovernmental Agreements between Member States and Third Countries in the Field of Energy Text with EEA Relevance." *Official Journal of the European Union* L 299 (27/10/2012): 13–17.

European Parliament and the Council. 1998. *Directive 98/30/EC Concerning Common Rules for the Internal Market in Natural Gas, 22 June*. Brussels: European Parliament and Council of the European Union.

European Parliament and the Council. 2003. *Directive 2003/55/EC Concerning Common Rules for the Internal Market in Natural Gas and Repealing Directive 98/30/EC, 26 June*. Brussels: European Parliament and Council of the European Union.

European Parliament and the Council. 2009. *Directive 2009/72/EC Concerning Common Rules for the Internal Market in Electricity and Repealing Directive 2003/54/EC (Text with EEA Relevance), 13 July*. Brussels: European Parliament and Council of the European Union.

European Parliament and the Council. 2010. "Regulation (EU) No 994/2010 of the European Parliament and of the Council of 20 October 2010 Concerning Measures to Safeguard Security of Gas Supply and Repealing Council Directive 2004/67/EC (Text with EEA Relevance)." *Official Journal of the European Union* L 295 (12.11.2010): 1–22.

Folketingstidende, A. 2017. *Lovforslag nr. L 43: Lov om ændring af lov om kontinentalsoklen*. Copenhagen: Folketinget.

FT. 2019. *Nord Stream 2 eyes way to curb EU oversight of $9.5bn pipeline*, March 14.

Goldthau, A., and N. Sitter. 2018. "Regulatory Power or Market Power Europe? Leadership and Models for External EU Energy Governance." In *New Political Economy of Energy in Europe: Power to Project, Power to Adapt*, edited by J. M. Godzimirski. Basingstoke: Palgrave.

Goldthau, A. 2018. *The Politics of Shale Gas in Eastern Europe. Energy Security, Contested Technologies and the Social Licence to Frack*. Cambridge: Cambridge University Press.

Goldthau, A., and N. Sitter. 2014. "A Liberal Actor in A Realist World? the Commission and the External Dimension of the Single Market for Energy." *Journal of European Public Policy* 21 (10): 1452–1472. doi:10.1080/13501763.2014.912251.

Goldthau, A., and N. Sitter. 2015a. *A Liberal Actor in A Realist World. The European Union Regulatory State and the Global Political Economy of Energy*. Oxford: Oxford University Press.

Goldthau, A., and N. Sitter. 2015b. "Soft Power with a Hard Edge: EU Policy Tools and Energy Security." *Review of International Political Economy* 22 (5): 941–965. doi:10.1080/09692290.2015.1008547.

Herranz-Surrallés, A. 2015. "An Emerging EU Energy Diplomacy? Discursive Shifts, Enduring Practices." *Journal of European Public Policy* 23 (9): 1386–1405. doi:10.1080/13501763.2015.1083044.

Högselius, P. 2012. *Red Gas. Russia and the Origins of European Energy Dependence*. Basingstoke: Palgrave Macmillan.

Judge, A., and T. Maltby. 2017. "European Energy Union? Caught between Securitisation and 'Riskification'." *European Journal of International Security* 2 (2): 179–202. doi:10.1017/eis.2017.3.

Kotek, P., A. Selei, and B. T. Tóth. 2017. *The Impact of The Construction of The Nord Stream 2 Gas Pipeline On Gas Prices and Competition*. Budapest: REKK Foundation.

Lang, K.-O., and K. Westphal. 2017. *Nord Stream 2 – A Political and Economic Contextualisation*. Berlin: Stiftung Wissenschaft und Politik. SWP Research Paper 2017/RP 03.

Liefferink, D., and M. S. Andersen. 1998. "Strategies of the 'Green' Member States in EU Environmental Policy-making." *Journal of European Public Policy* 5 (2): 254–270. doi:10.1080/135017698343974.

Lijphart, A. 1977. *Democracy in Plural Societies: A Comparative Exploration*. New Haven: Yale University Press.

Lodge, M. 2008. "Regulation, the Regulatory State and European Politics." *West European Politics* 31 (1/2): 280–301. doi:10.1080/01402380701835074.

Loskot-Strachota, A. 2016. *Nord Stream 2: Policy Dilemmas and the Future of EU Gas Market*. Oslo: NUPI.

Lukes, S. 1974. *Power: A Radical View*. London: MacMillan Press.

Majone, G. 1994. "The Rise of the Regulatory State in Europe." *West European Politics* 17 (3): 77–101. doi:10.1080/01402389408425031.

Moravcsik, A. 1993. "Preferences and Power in the European Community: A Liberal Intergovernmentalist Approach." *Journal of Common Market Studies* 31 (4): 473–524. doi:10.1111/j.1468-5965.1993.tb00477.x.

Natorski, M., and A. Herranz-Surrallés. 2008. "Securitizing Moves to Nowhere? The Framing of the European Union's Energy Policy." *Journal of Contemporary European Research* 4 (2): 70–89.

Nosko, A., and M. Mišík. 2017. "No United Front: The Political Economy of Energy in Central and Eastern Europe." In *Energy Union. Europe's New Liberal Mercantilism?* edited by S. Andersen, A. Goldthau, and N. Sitter, 201–222. Basingstoke: Palgrave Macmillan.

Puetter, U. 2014. *The European Council and the Council: New Intergovernmentalism and Institutional Change*. Oxford: Oxford University Press.

Schattschneider, E. E. 1960. *The Semisovereign People: A Realist's View of Democracy in America*. New York: Holt, Rinehart and Winston.

Schmidt-Felzmann, A. 2020. "Gazprom's Nord Stream 2 and Diffuse Authority in The Eu: Managing Authority Challenges Eegarding Russian Gas Supplies Through The Baltic Sea." *Journal of European Integration* 42 (01).

Spencer, M. E. 1970. "Weber on Legitimate Norms and Authority." *The British Journal of Sociology* 21 (2): 123–134. doi:10.2307/588403.

Surrallés, A. H., I. Solorio, and J. Fairbrass. 2020. "Renegotiating Authority in The Energy Union: A Framework for Analysis." *Journal of European Integration* 42 (01).

Szulecki, K., ed. 2017. *Energy Security in Europe. Divergent Perceptions and Policy Challenges.* London: Palgrave.

Szulecki, K., S. Fischer, A. T. Gullberg, and O. Sartor. 2016. "Shaping the '"energy Union': Between National Positions and Governance Innovation in EU Energy and Climate Policy." *Climate Policy* 16 (5): 548–567. doi:10.1080/14693062.2015.1135100.

Talus, K. 2011. "Long-term Natural Gas Contracts and Antitrust Law in the European Union and the United States." *The Journal of World Energy Law & Business* 4 (3): 260–315i. doi:10.1093/jwelb/jwr015.

Weber, M. 1922. *Wirtschaft und Gesellschaft.* Tübingen: Mohr.

Weber, M., R. Bütikofer, N. Hirsch, E. Brok, N. Röttgen, O. Krischer, and M. Link. 2018. "Nord Stream 2 Schadet Europa." *Frankfurter Allgemeine Zeitung*, February 20.

Gazprom's Nord Stream 2 and diffuse authority in the EU: managing authority challenges regarding Russian gas supplies through the Baltic Sea

Anke Schmidt-Felzmann (iD)

ABSTRACT

This article addresses a neglected question about the effects of dispersed authority in the European Union (EU) on the EU's ability to manage external contestations. It investigates authority challenges from two major external actors in the energy, and specifically the gas sector, and scrutinizes the management of authority conflicts at local sites in different EU states in the Baltic Sea region. The second Russian transboundary submarine gas infrastructure project *Nord Stream 2* (NS2), owned and managed by a whole-owned subsidiary of the Russian state-controlled energy giant *Gazprom*, serves as an illustrative case study. It reveals how multiple authority conflicts at different sites and levels, and the challenges from both the Russian Federation (Russia) and the United States of America (USA), played out, and why formal adjudication has been a primary tool for the EU and several of its member states to assert their authority and manage contestations at the local, national and EU level.

Introduction

For more than three decades, the European Union (EU) has struggled to establish its authority in the energy sector (McGowan 2011). External actors, notably Russia, have exploited overlapping competences and stoked authority conflicts between and within the member states and EU institutions. The market distorting activities of Gazprom, one of Russia's state-controlled energy giants operating in the EU, have challenged the EU's ability to assert its authority. The European Commission's efforts to create, with the support of the Court of Justice of the European Union (CJEU), a level playing field and ensure an open competition between gas market actors across the EU, have resulted in the imposition of multiple corrective measures. Gazprom has repeatedly been forced to revise its supply contracts with European companies (Schmidt-Felzmann 2019b). This article focuses on a case that illustrates how the dispersion of authority within the EU affects its ability to manage external contestations in the energy realm. It investigates authority challenges from Russia (Gazprom and its subsidiaries) and the USA, two major external actors. It scrutinizes how authority conflicts are managed, considering also local sites in different EU states, in addition to national, EU level and international contestations. The major transboundary (sub)marine gas infrastructure project, composed of *Nord Stream* (NS1) and *Nord Stream 2* (NS2), each consisting of two parallel pipelines running from Russia across the Baltic Sea to Germany, serves as an illustrative case.

NS1 – conceived of during the Soviet period, prepared in the 1990s, and implemented during the 2000s – was the source of major conflicts with Russia, and heavily contested within and between the affected EU member states (EP 2007), even more so following the Russian-Georgian war in 2008, which drew attention to Russia's geostrategic ambitions in its neighbourhood. The United States of America (USA; US) supported political actors in the EU that opposed the NS1 construction. Its contestation of the Russian gas project to Germany was a continuation of a principled US opposition to the Soviet Union's gas supply projects to West Germany which were implemented successively from the 1960s (Högselius 2013) despite US objections. NS1 received EU infrastructure support (EP 2007). Since 2012 it delivers Russian gas through the first set of submarine pipelines to Germany. The OPAL connector (*Ostsee-Pipeline-Anbindungsleitung*), built to feed the Russian gas from Lubmin into the Central and East European distribution network, came to benefit from an exemption related to the obligations imposed by the EU's 'Third Energy Package' of 2009. Proposed by the German regulatory authority (*Bundesnetzagentur*) and negotiated with the European Commission (Bundesnetzagentur 2019), the exemption from which NS1 deliveries could benefit, became heavily contested by Poland, with the support of Lithuania (CJEU 2019, T 883/16). It has consequences for the role of Poland and Ukraine as transit states, and for the profitability of Russian supplies of gas through the Baltic Sea. The patterns of contestation in the EU with/against Russia, and the USA in a secondary role, were repeated in a modified manner with NS2. Launched in 2015, NS2 quickly turned into a bone of contention between Russia and the affected EU states. The effects on Ukraine, as the main transit state for Russian gas to the EU, was at the centre of attention. Its commercial losses and Russia's armed aggression against it, served opponents as central arguments; supporters of NS1 were swayed, swerving towards seeking a ban on NS2.

The controversy about NS2 reveals how diffuse authority affects the EU, namely the European Commission's ability to manage authority challenges at multiple levels and sites. As the EU's *guardian of the treaties*, the European Commission holds a central position in authority battles about external energy supplies (Mayer 2008; McGowan 2011), while the CJEU ensures that the institutions and members alike fulfill their duties (e.g. CJEU 2019). This investigation focuses on different axes of authority contestation within and beyond the EU's energy acquis, distinguishing between (a) the *sovereignty-based* contestation of legal competences (horizontally and vertically) and (b) the *substance-based* contestation (horizontally). It examines how different concerns – (i) commercial interests, (ii) security concerns and (iii) environmental protection – are prioritized and used as leverage, and with what effect. The analysis of the different levels (local, national, EU and international) and arenas or sites (geographic and/or issue-specific) reveals a range of interlocking challenges, conflicts of interests and questions of legal interpretation, cutting across sectoral divisions (environment, energy, economy, national security) and legal boundaries (national, EU, international). The study is based on an extensive qualitative analysis of primary and secondary sources, background discussions with decision-makers, diplomats and subject experts on NS2 (2015–2019), complemented by an analysis of published records and media reports of official debates in the EU on NS1 (2005 to 2011).

The article is structured as follows: the first part locates the analysis in the academic debates and outlines its contribution. The second part looks at three axes of authority contestation to examine how the member states have managed challenges from NS2 in their (1) assertion of authority towards challenges that are channelled through the local level – *without* involving the EU; (2) assertion of authority *in opposition to* the EU; (3) assertion of authority *with* the EU. The third part scrutinizes how the EU itself has managed (1) the primary authority challenges from (i) Russia and (ii) Gazprom's subsidiary NS2 and (2) the secondary challenges from the USA. The final part draws conclusions from NS2 about the EU's management of diffuse and contested authority from external actors.

Exploring authority contestation at multiple levels

A significant body of research has examined the drivers of Russian gas supply policies to EU member states (Bilgin 2011; Sharples 2016) and the effects of Gazprom and the Russian gas sector on European energy supplies (Henderson and Moe 2016; Locatelli 2014; Boussena and Locatelli 2017; Finon and Locatelli 2008). Another cluster of research has examined the EU's 'power' in the energy sector, specifically regarding Russia (Goldthau and Sitter 2014, 2015; Schmidt-Felzmann 2019b, 2019c) and the regulatory effects of the EU's gas directives and their legal application (Romanova 2016; Roth 2011). The effects of EU energy law on the member states (Andersen and Sitter 2015) and the European Commission's engagement with energy companies (Stoddard 2017) and legal aspects of NS1 (Langlet 2014a, 2014b) have received increasing attention. In addition, Lee et al. (2008) examined the legal consequences of NS1 for the EU institutions, Marhold (2012) investigated NS1 in the context of Russia's exit from the Energy Charter Treaty (ECT) and Talus and Wüstenberg (2017) elaborated upon ramifications of international frameworks for NS1 and NS2.

The environmental, economic and governance consequences of NS1 (Karm 2008; Lidskog and Elander 2012; Lidskog, Uggla and Soneryd 2011) and the competing interests and dilemmas in the *Environmental Impact Assessment*(EIA) for governments dealing with NS1 (Koivurova and Pölönen 2010) attracted the attention of local researchers. International obligations with the EIA have been explored in terms of its function as a governance tool in the Baltic Sea (Ifflander and Soneryd 2014). The framing of problems and ways in which environmental criteria were addressed in the assessment of NS1 has engaged Swedish researchers (Lidskog and Elander 2012). Another cluster explored the positions of Poland and Germany (Bouzarovski and Konieczny 2010; Wojcieszak 2017), and of Sweden on NS1, including the local authorities' role (Edberg, Fransson, and Elander 2017; Sayuli, Elander, and Lidskog 2011). Research from Norway and Finland has investigated Russia's geo-economic and geo-strategic driving forces in its neighbourhood, concluding that foreign policy, rather than purely commercial interests, underpin NS1 and NS2 (Godzimirski 2011; Vihma and Wigell 2016).

This article moves beyond existing studies by focusing on the EU's and member states' management of authority conflicts. It analyses how, and where, authority claims that are *sovereignty-based* and those that are *substance-based* are employed in challenges and counter-responses to NS2. Sovereignty-based authority contestation refers to conflicts over where the legal competence to take a decision is held, that is *who exercises* authority, while *substance-based* refers to *how* the competent actor(s) claim their authority (Herranz-Surralles, Solorio Sandoval, and Fairbrass 2020). The external authority challenges in the NS2 development (permitting and implementation) are identified, and it is then examined how conflicts are managed. NS2 has engaged different governmental, sub-national and non-governmental actors in the Baltic Sea area. The commercial and logistics contracts and permits, without which NS2 could not be built, turn the jurisdiction over sites required for its construction into key battlegrounds. Against this background, this article seeks answers to when, why and in how far the *management* of authority conflicts relies on *formal adjudication,* or on *political strategies* (Herranz-Surralles, Solorio Sandoval, and Fairbrass 2020).

The member states' management of authority challenges from NS2

International legal frameworks govern the permitting and implementation process at the national level. The *United Nations Convention on the Law of the Sea* (UNCLOS), together with the *Convention on the Protection of the Marine Environment of the Baltic Sea* (Helsinki Convention), and the *Convention on Environmental Impact Assessment in a Transboundary Context* (Espoo Convention), form the body of international law that applies to the permitting decisions on NS2. UNCLOS (1982) stipulates rules to be followed when wishing to lay submarine pipes on the continental shelf of another country. Objections by affected coastal states can, in principle, only be motivated by environmental concerns, such as the impact on the natural habitat. Because UNCLOS builds on the UN Charter's principles, arguably the violations of fundamental principles of state conduct, notably of national sovereignty and territorial integrity, constitute such severe breaches of the basic principles enshrined in the UN Charter that they could provide grounds for contesting Russian claims to an entitlement to lay subsea pipes in the Baltic Sea after the gross violations that the Russian state committed with the annexation of Crimea (Riley 2016).

According to the Espoo Convention, national authorities have to base their decision on permits for commercial activities upon EIAs. National rules and procedures determine the extent to which other actors can be heard in the permitting process. The NS2 permitting involved both national and municipal authorities. The route for NS2 was to run, in parallel to NS1, through the Finnish, Swedish, Danish and German *Exclusive Economic Zone*s (EEZ) and the territorial waters of Germany, where NS2 – like NS1 – would come onshore at Lubmin. In contrast to NS1, NS2 considered that the optimal route around the Danish island of Bornholm would run through Danish *territorial* waters. Permits were required not just for the pipeline itself, but also for the associated logistics work in the vicinity of the route. In Finland, Sweden and Germany, applications for the required permits were submitted in 2016 and granted in 2018. The Finnish approval was granted without any serious challenges. Both the government and the Regional State Administrative Agency for Southern Finland granted the permits for the construction in April 2018.

The assertion of authority against challenges channelled through the local level

The extensive Swedish referral process (*remissväsendet*) allows state and non-state actors alike to participate in the permitting process, and make submissions. NS2 applied for the permit in September 2016 (Government Offices of Sweden 2016a), which it obtained in June 2018 (Government Offices of Sweden 2018). Within the Government Offices (*Regeringskansliet*), the Ministry for Enterprise and Innovation (*Näringsdepartementet*) was the responsible department (for details on the process, see Schmidt-Felzmann

2019a). Between 2016 and 2018, the Swedish NS1 and NS2 representative, Lars O. Grönstedt, and the Russian Ambassador expressed in various public statements their expectation that the Swedish government will grant the permit and support NS2. The opposition parties, meanwhile, were strongly critical of how NS2 was being handled, accusing the government of inertia and of jeopardizing Sweden's national security unless it refused the permit and 'stopped' NS2.

In 2016, the principle of local self-government (*det kommunala självstyre*) turned into a key battleground, as NS2 opponents chastised the government for its unwillingness – or inability – to interfere in two municipalities' decisions on the acceptance of business deals with the Gazprom subsidiary (Schmidt-Felzmann 2019a). Both a major port in the Municipality of Karlshamn and a small harbour on the East coast of Gotland had received commercial offers from NS2 to serve as logistics sites. The Port of Karlshamn in Blekinge is whole-owned by the Municipality and one of Sweden's largest Baltic Sea ports. Blekinge Bay is also an important exercise area for the Swedish Navy and the Air Force's *Blekinge Wing* F17. Together with another smaller harbour (Öxelösund), Karlshamn had featured among the possible storage sites for NS1, a decade earlier. In 2016, Karlshamn purchased 200,000 square metres storage space at Stilleryd harbour to increase the Port's capacity for NS2. For Karlshamn, it was of paramount importance to recover these investments. Slite harbour on Gotland is whole-owned by the regional authority governing the island. It was used as logistics site for NS1, and is located near the NS2 route. Slite and Gotland were, at the time, financially hard pressed. The new national Defence Bill of 2015 had, however, prescribed the reinstallation of a permanent military presence on Gotland by 2017, and in September 2016, troops were rotated onto the island.

In December 2016, responding to a deteriorating security environment with Russia, representatives of Karlshamn and Gotland were called to an urgent meeting with the Defence and Foreign Ministers and the Swedish security and intelligence services. The government's objections to the Municipalities' deals with NS2 were motivated by anticipated negative consequences for Sweden's national security (Government Offices of Sweden 2016b; SVT 2016). After the meeting, the Chairman of Gotland's Regional Assembly conceded that these concerns weighed so strongly that Slite would withdraw from the NS2 deal. At a follow-up meeting in January 2017, just hours before the German Chancellor arrived in Stockholm for a state visit, Karlshamn announced that the NS2 deal would, despite these concerns, go ahead. The Ministers underlined that while NS2 was deemed harmful to Sweden's security, the local '*självstyre*', and Swedish obligations under UNCLOS, tied the government's hands. Security measures would be put in place to handle the risks associated with NS2 at Stilleryd harbour and in Blekinge Bay (Ahlborg and Danielson 2016). In the end, Karlshamn's commercial interests trumped Swedish hard security concerns. NS2 and the Russian Ambassador welcomed Karlshamn's decision and underlined its commercial nature and important contribution to the EU's energy supply (Nord Stream 2 2017). The Ambassador, and NS2, had earlier accused the Swedish government of 'interfering' and politicizing the 'purely commercial' pipeline project. Research into NS1 had found that environmental concerns were downgraded to the benefit of Russian and German economic interests (Lidskog and Elander 2012). During the permitting process for NS2, environmental arguments played hardly any role; the conflict was concentrated around commercial versus (hard) security arguments.

In April 2017, a similar authority conflict emerged from NS2 in Latvia, when the Mayor of Ventspils and Chairman of the Port, Aivars Lembergs[1], announced that the Noord Natie Ventspils Terminal would enter a deal with NS2 and act as a logistics and cargo handling site (Baltic News Network 2017a, 2017b, 2017d). The Latvian government had a formal veto, since 50% of the Port's Board is controlled by the state. It used these powers to prevent a participation in NS2 (Freeport of Ventspils n.d.). In response, the Port authorities, supported by the Baltic Association for Transport and Logistics (BATL), and endorsed by the Mayor, submitted a compensation request to the government for lost cargo transport services, cargo storage services, unused supply of technological elements (territory, machinery investments) and lost profits from railway transportation services for the Baltic Express (Baltic News Network 2017c, 2017d). From its position of authority, the Latvian government's objection on hard security grounds trumped the Municipalities' commercial arguments.

A similar commercial opportunity on Denmark's island of Bornholm at Rønne havn (Port of Rönne) was rejected out of hand. When it became clear that its location might help it clinch a deal with NS2, the Mayor announced that Rønne would not go against the express wishes of the Danish government (TV2/Bornholm 2017). NS2 submitted three applications to obtain its permit from Denmark, each for an alternative path, running (1) South of Bornholm through *territorial waters*, (2) North of Bornholm through the EEZ and (3) South of Bornholm through the EEZ, through a maritime zone embroiled in a Danish-Polish border dispute, which was finally settled in November 2018. Denmark's Parliament (*Folketing*) proceeded in 2017 to clarify the state's authority in its territorial waters, affecting NS2 Route 1. The competent Danish Energy Agency (*Energistyrelsen*) was still assessing the application for Route 1, which meant that the amendments could impact NS2. Therefore, NS2 threatened Danish decision-makers of legal consequences deriving from World Trade Organization (WTO) and ECT obligations, unless they dropped the amendments (Law Firm Bech-Bruun 2017, 9 August). The legal claims were supported by an NS2-commissioned memorandum (Schill 2017). Despite these threats, the legislative amendment to the *Danish Law on the Continental Shelf* swiftly passed parliamentary scrutiny, empowering the government as of 1 January 2018 to take national security and defence concerns into consideration in any permitting procedures that concerned its territorial waters (Folketing 2017).

The national assertion of authority in opposition to the EU

In Germany, the local authorities of Mecklenburg West-Pomerania (MWP) and the national government endorsed NS2, as they did already with NS1 (EP 2007). Already in 2015, the Minister for Energy and the Environment had made clear in Moscow that the German government would seek to '*prevent external meddling*'. The Russian presidential administration cited and published the German Minister's pledge of support:

> "What's most important [...] is that we strive to ensure that all this remains under the competence of the German authorities, [...] if we can do this, then opportunities for external meddling will be limited. [...] What's most important is for German agencies to maintain authority over settling these issues. And then, we will limit the possibility of political interference [...] (President of Russia 2015).

In 2018, the Prime Minister of MWP similarly asserted the federal states' decision-making authority on NS2 (NDR1 Radio MV 2018). These assertions were anchored in a claim that the *Third Gas Directive* was not applicable to German coastal waters (BMWI 2018), and that decisions on NS2 were a national prerogative – a claim that others refuted (Riley 2016).

Two German authorities had to grant permits to NS2: the Stralsund Mining Authority (*Bergamt Stralsund*) and the Federal Maritime and Hydrographic Agency (*Bundesamt für Seeschiffahrt und Hydrographie, BSH*) in Hamburg. The *Bergamt's* authority regarding the construction of NS2 in the coastal (German territorial) waters of Mecklenburg WP derives from the national Energy Industry Law (*Energiewirtschaftsgesetz, EnWG*). The permit was granted in January 2018. In its motivation, priority was given to German commercial interests above any environmental concerns (Bergamt Stralsund 2018). The Federal Mining Law (*Bundesberggesetz*, BBergG) grants the BSH the authority to approve the NS2 route in the German EEZ. It granted the permit in March 2018[2].

A German NGO, the Nature and Biodiversity Conservation Union (*Naturschutzbund (NABU) Deutschland*) challenged the construction permit on environmental grounds (NDR 2018a). It went to the Higher Administrative Court (*Oberverwaltungsgericht (OVG)* in Greifswald, which rejected NABU's appeal for an injunction to stop the NS2 work until its concerns about the environmental impact on the territory of MWP and its coastal region had been properly considered. NABU submitted also an urgent appeal (*Eilantrag*) to the Higher Regional Court (*Oberlandesgericht* (OLG) in Rostock, and to Germany's highest court, the Federal Constitutional Court (*Bundesverfassungsgericht)* in Karlsruhe to obtain an injunction. Its appeals at all administrative levels remained unsuccessful (NDR 2018d). NABU criticized the Courts' refusal to impose an injunction on the constructon work until its legal cases against NS2 had been settled, and protested that in the absence of an independent impact assessment, 'the economic interests of a large company [i.e. NS2/Gazprom] were ranked higher than nature conservation' (NDR 2018e).

As with NS1, the commercial interests of the local and national authorities may have acted as a further motivation to push ahead with NS2, at the possible expense of the environment (Schmidt-Felzmann 2019a). Among others, municipal taxes paid by NS1 to Lubmin are substantial, an aspect of importance in a region that is economically hard pressed (NDR 2018b). When in May 2018, just a week after the construction work had started, a sudden environmental contamination on the German coast was traced back to the preparatory dredging (NDR 2018e), NS2 suspended its work and pledged to pay for the clean-up. Despite the incident, the environmental organizations' and local Green Party's contestation of the federal state and national government's decisions were entirely unsuccessful. The Bergamt Stralsund emphasized in its motivation for granting the final permit that the speedy implementation ('*sofortiger Vollzug*') of NS2 was in the German public interest (Bergamt Bergamt Stralsund 2018, 11). The BSH followed in its own assessment of the NS2 route, the same argumentation, and so did the local and national Courts.

The national assertion of authority with the EU

Several member states severely criticized the German position. Already NS1 gave rise to intense challenges from Poland and Sweden, among others (Lidskog and Elander 2012;

Sayuli; Fransson, Elander, and Lidskog 2011; Vihma and Wigell 2016; Wojcieszak 2017). In its principled opposition to NS2, Poland pursued a formal adjudication by national and European courts. The Polish state-controlled energy company *Polskie Górnictwo Naftowe i Gazownictwo* (PGNiG), together with the national regulator, used the EU's energy market and competition rules to challenge NS2, while simultaneously contesting the conditions under which NS1 could operate (Bundesnetzagentur 2019). In mid-2016, the Polish *Office of Competition and Consumer Protection* (UOKiK) raised objections against Gazprom's NS2 plans, when it became clear that the Russian company (controlling 50–51%) planned to form a joint venture with five European companies as shareholders (together 49–50%). UOKiK contested these plans, as it would create 'an excessive market concentration' of these companies in Poland (UOKiK 2016). In response to UOKiK's plans to launch legal proceedings against them, the five companies dropped out, and in August 2016, NS2 became a whole-owned (100%) subsidiary of Gazprom.

UOKiK referred to the EU's anti-trust rules again in April 2017, when Gazprom announced a co-financing agreement. The five companies pledged to financially support NS2, without becoming shareholders. UOKiK argued that this arrangement was – just like the joint venture – in breach of anti-monopolist rules as it would result in a *de facto* market concentration (UOKiK 2018). Poland challenged also an interim decision on OPAL that allowed Gazprom to increase its gas supply through NS1 (Bundesnetzagentur 2019). The ECJ ruled in September 2019 in favour of Poland, underlining in its judgement the principle of energy solidarity enshrined in EU law (CJEU 2019, T883/16), and which Poland had more than a decade earlier managed to make legally binding with the *Lisbon Treaty* (Roth 2011). Ukraine received, at the same time, support from the European Commission in trilateral negotiations with Russia on future gas transit through its territory (European Commission 2014).

The Energy Ministers of Denmark and Sweden had already earlier requested, in a joint letter, that the NS2 conflicts be managed at the EU level, in line with the Energy Union ambitions and solidarity principles. Supported by other EU states (SVT 2017) as well as the EP (2018, 2019), they countered German claims of the benefits of NS2 for European energy supply security. NS2 had used these arguments to market and defend the project since its launch in 2015. The German Green Party challenged the federal government in a series of parliamentary questions (*Kleine Anfrage, Grosse Anfrage*, e.g. BMWI 2016) of uncritically following Gazprom's own projections (NDR 2018c, 2018f, 2018g). The estimates of future energy demands were challenged also by the EP (2015, 2018, 2019).

The EU's management of external authority challenges regarding Nord Stream 2

The management of Russian authority challenges

The EU's competences on NS2 remained controversial and the European Commission concluded that the Council should empower it to negotiate an agreement with Russia on NS2 to ensure legal certainty for investors while moderating negative consequences for the EU's gas market. In June 2017, the European Commission presented its draft negotiating mandate.[3] In the proposed negotiations with Russia it aimed to ensure that NS2 'operates with an appropriate degree of regulatory oversight, in line with key principles of international and EU energy law' (European Commission 2017a). Germany, Austria and France, backed by a couple of other EU states rejected the mandate. Consequently, the modification of the Gas Directive (2009/73/EC) became a priority for the European Commission to ensure legal clarity for *all* new pipelines entering the EU, a priority underlined also in its President's *State of the Union* speech of September 2017 (European Commission 2017b). The amended Directive would apply at the very least to 'the EU-part' of NS2, while the non-EU part would either be subject to EU law *'by territorial extension'* or be governed by an EU-Russia agreement relating to the whole pipeline (Dudek and Piebalgs 2017). The legislative proposal challenged the German assertion that it would determine "on its own" the rules that would govern the operation of the new pipelines.

The EU's energy market rules, especially the *Third Energy Package*, have a direct bearing on NS2 in terms of its ownership and operation, once completed (Riley 2010, 2018). The Russian state pursued several challenges to the gas market directive (2009/73/EC), both at the WTO (European Commission 2018) and within the ECT (Dralle 2018, 25–26; Marhold 2012). At the WTO, Russia submitted its complaint in April 2014. The dispute settlement panel confirmed in its report of August 2018 the EU's right to enforce its market rules on Gazprom with the 'Third Package' (Schmidt-Felzmann 2019c). A verdict on the EU's and Russia's appeals to the appellate body's findings became delayed due to a work overload and the legal complexity of the case, as notified to the parties on 21 November 2018.

In the EP, a majority had, from the beginning, opposed NS2 (EP 2015, 2018), like it earlier challenged even NS1 (EP 2007). It demanded that NS2 decisions be decided on an equal footing with the Council, and emphasized that the EU must assert its authority:

> it is in the Commission's and Council's hands to ensure that projects which contradict the principles of the Energy Union, such as the expansion of Nord Stream, will not be granted any EU support, financial or other, including derogations from the 3rd Energy Package. (EP 2015).

In February 2019, on the basis of a French-German proposal to change certain stipulations in the amendments to the gas directive concerning the delineation of the national and EU authority on gas pipelines with foreign states, an agreement was finally was reached in the Council (European Commission 2019a): EU rules will apply only to the territory and territorial waters of the state where the first entry point is located, and *not*, as proposed by the European Commission, to the EU as a whole. Consequently, the German *Bundesnetzagentur* retained a decisive say over how the EU's gas market rules are to apply to NS2 (Bundesnetzagentur 2019; BMWI 2016), albeit under the oversight of the European Commission. The EP, in co-decision with the Council, approved the amendments and underlined that their adoption asserted the EU's authority: 'From now on [. . . even] Nord Stream 2, will have to abide by EU rules: third-party access, ownership unbundling, non-discriminatory tariffs and transparency.' (EP 2019).

In 2017, during its battle against the Folketing's legislative amendments, a challenge to the EU through the ECT was brought into play by several NS2 consultants and legal representatives (Bech-Bruun 2017; Schill 2017; Talus 2017). Their claim was that such legislative amendments were undermining predictability, and therefore in breach of the ECT's and WTO's investment protection rules, turning NS2 into a victim of discriminatory treatment (first by Denmark, then the EU). In April 2019, NS2 formally contested the legality of the new Directive (2019/692). It conceded that NS2 was not completed, nor operational, but must be treated either as 'completed' or be granted a derogation (NS2 2019a). The NS2 construction had progressed towards Bornholm (Sjöfartsverket 2019, UFS 17 and 24 October), but by late November 2019, the pipelines remained physically 'un-completed'. The last required construction permit from Denmark had been granted (for Route 3) in late October, taking into account the Armed Forces' concerns (Energistyrelsen 2019a, 2019b). With the completion of NS2 still pending, the authority battles from the external challengers regarding both the pipeline construction and the regulation of its operation, continued – *with* and *against* the EU – into 2020.

In July 2019, supported by another commissioned memorandum (Herbert Smith Freehills 2019), NS2 reiterated its demand that the pipelines be treated by the European Commission as 'completed before 23 May 2019', the date of the entry into force of the amended EU Directive (NS2 2019b). NS2 submitted its demand to the CJEU to annul the Directive (NS2 2019c) and in September 2019, it submitted its complaint under the ECT. As Eichberger (2019) emphasized, while '[i]n the Yukos Saga, the Russian Federation emphatically opposed the provisional application of the dispute settlement provisions of the ECT [,] Gazprom's subsidiary [was now] using just these dispute settlement mechanisms' against the EU (see also Riley 2010). The Russian state had suffered a series of (at least partly) successful challenges from the former Yukos shareholders at The Hague Court of Arbitration, until it achieved a (partial) victory at The Hague District Court (Voon and Mitchell 2017; Alvarez Sanz 2017). The European Commission (2019b), in its response to NS2's threat of ECT proceed-ings, clarified in July that the authority for granting a derogation rested with *Bundesnetzagentur*. In November 2019, NS2 received support from German lawmakers who, in the legal text to transpose the Directive, increased the wriggle room for Bundesnetzagentur to take 'the special circumstances of the individual case' ('*alle Umstände des Einzelfalls*') into consideration in determining the status of new pipelines as 'completed before [Directive] 2019/692 entered into force' (Handelsblatt 2019). The trans-position and its implementation can be contested, if they contradict EU rules.

The management of US authority challenges

In 2017, challenges from the USA to NS2 opened up a new battlefront. The US Congress targeted NS2 in various Resolutions of the *House of Representatives* and *Senate,* and with a series of draft Bills on Russia that had the stated aim of imposing costs on Russia for its aggression against Ukraine, and to strengthen the (energy) security of US allies in Europe. When US Congress moved ahead with a larger legislative act on the basis of which it threatened to impose sanctions against European companies participating in NS2, to stop its construction and prevent its completion, the Austrian and German governments, in a joint statement, rejected the USA's 'extraterritorial sanctions' (German Foreign Ministry 2017). The German Minister and Austrian Chancellor claimed they represented the EU as a whole which they insisted, needed to assert its authority against the USA (German Foreign Ministry 2017). The European Commission President, and the French government, quickly endorsed this assertion. The rejection of the envisaged sanctions on Russia as 'US meddling in European affairs' were backed by allegations that the main driving force of Congress and the US President were prospective commercial gains for privately owned US-based entities producing LNG, whose deliveries to Europe could potentially replace (some) of the Russian pipeline gas, if the construction of NS2 was stopped.

Although the *Countering America's Adversaries Through Sanctions Act* (CAATSA), was adopted (US Congress 2017b), incorporating the Senate's *Countering Russian Influence in Europe and Eurasia Act* (US Congress 2017a), the envisaged measures against NS2 under CAATSA were halted. Germany and France, together with representatives from the European Commission's DG Trade, had threaten US lawmakers and the US President with the imposition of European countersanctions, unless NS2 and European companies were exempted from such measures. CAATSA served meanwhile as the legal basis for a number of sanctions targeting the Russian intelligence and defence sector (United States Code 2019). In 2018 and 2019, several more draft Bills and bipartisan Resolutions were developed in Congress, expressing the strong US parliamentary opposition to the completion of NS2, and declaring the explicit purpose of imposing sanctions on those involved in NS2 operations (US Congress 2018a; 2019a; 2019b; 2019c; 2019e). Taking into consideration the USA's sovereign authority to impose sanctions, combined with its status as a major trade actor and financial market, the legal instruments of CAATSA sec. 235 (US Congress 2019f) enabled it to delay the completion and make the operation of NS2 more difficult.

In November 2019, it was announced that the US *National Defense Authorization Act (NDAA) 2020* (US Congress 2019g) would include sanctions against NS2 through the incorporation of the *Protecting Europe's Energy Security Act* (US Congress 2019d). In December 2019, the inclusion of the bipartisan bill (US Congress 2019d) in the NDAA 2020 was confirmed by one of its initiators (Cruz 2019). Since NS2 remained physically 'uncompleted' at this point, the NDAA 2020 made it possible for the US, with strong bipartisan parliamentary support, to exert its authority in Europe, and against the Russian gas infrastructure project, by targeting European (and other) companies and individuals 'financially or with services enabling and facilitating the construction, maintenance or expansion of Russian energy export pipelines' (US Congress 2019f). This instrument was, upon the entry into force of the NDAA on 21 December 2019, deployed to deter subcontractors and other commercial actors needed for NS2's completion in the Baltic Sea, and for its operation in the EU.[4]

Conclusions

The NS2 case illustrates how dispersed authority provides opportunities for external actors to contest the EU's authority in a specific realm (gas, pipeline construction), across hierarchical levels and national boundaries. The role played by local authorities deserves closer attention, as well as how different local, national, EU and international frameworks are used both to challenge, and to assert, authority claims. In the management of authority conflicts, the domestic distribution of authority at the specific geographic site of contestation can help explain when and why legal adjudication, rather than political strategies, are preferred. NS2 puts a spotlight on the significant role in the EU of *formal adjudication* in the management of conflicts with Gazprom, and its subsidiary NS2. Despite their prevalence in other realms, political management strategies, as described by Herranz-Surralles, Solorio Sandoval, and Fairbrass (2020), have served here chiefly to counter challenges posed by the USA's (threats of) sanctions.

Hard security concerns have motivated national governments (Sweden, Denmark, Latvia) to empower the EU on energy, and to assert their own authority over local governments, while economic interests motivated local authorities (Karlshamn, Gotland,

Ventspils) to support NS2. In the case of Germany, the local and federal governments supported the external challenger (NS2, Russia, and Gazprom via OPAL), in opposition to the EU (and the USA). The use of political strategies towards the USA, and by Sweden towards Gotland, indicates that the choice of management strategy may be dependent on the bargaining power of the entity (EU, national or local government) that is challenged, and on whether it plays a primary, or a secondary role in the conflict.

The controversy surrounding the amendments of 2019 to the EU's Gas Market Directive demonstrates both the substantial national support for an empowerment of the EU, and the persistent unwillingness to delegate authority. Since compliance with other international obligations remains under the purview of the contracting states, so that the EU's own rules come into play only *after* NS2 enters into operation, the EU's ability to manage conflicts associated with the planning and permitting phase was clearly limited. In this regard, the study into NS2 underlines the constraints under which the EU operates as a global actor more generally speaking, and especially when dealing with Russia (Schmidt-Felzmann 2019c). Conflicting interests within the EU significantly reduce its ability to manage conflicts arising from external challenge(r)s. The US-German disputes surrounding the German defence of the NS2 project illustrate the particular difficulties that the Union faces when member states contest its authority in conjunction with a powerful external actor, against another powerful global actor. Nevertheless, the relocation of authority challenges from the EU to the WTO and ECT modifies these conclusions somewhat. While the settlement of conflicts in an arena that lies outside the EU's own legal purview poses additional challenges, the assertion of the EU's claims at the ECT (and at the WTO) would certainly empower it on its own turf, both in relation to NS2/Gazprom and Russia, and vis-à-vis those member states that have themselves contested its authority.

Notes

1. Aivars Lembergs, his wife and two children were placed on the USA's *Specially Designated Nationals* (SDN) sanctions list on 10 December 2019 (for details, see US Department of State 2019).
2. This permit was modified with a decision of 23 December 2019 by the BSH which approved NS2's application of 23 September 2019 for a permit to undertake the construction work for a 16.5km stretch of the pipeline in the German EEZ in early 2020. The NS2 request was submitted in response to delays caused by the permitting process in Denmark. The approval of the "winter permit" was contested by environmental protection NGOs.
3. The draft negotiating mandate of 9 June 2017 – an EU-restricted document (European Commission 2017c) – was made public by Politico.eu: European Commission (2017c) Recommendation for a Council Decision authorizing the opening of negotiations on an agreement between the European Union and the Russian Federation on the operation of the Nord Stream 2 pipeline, at http://www.politico.eu/wp-content/uploads/2017/07/NS-Draft-Mandate.pdf.
4. On 21 December 2019, the Swiss-registered pipelaying firm Allseas suspended its construction work on NS2, giving rise to speculations about whether or not, and if so when, NS2 may be completed and become operational despite the US sanctions - at at what cost.

Acknowledgments

The author thanks the anonymous reviewers for helpful suggestions and Jenny Fairbrass, Anna Herranz-Surrallés and Israel Solorio Sandoval for the energy that they have put into the organization of this Special Issue, as well as the UACES workshop on "Diffuse Authority and Contestation in the EU Energy Transition" in April 2018 at the University of Maastricht, and the many conference panels and energizing activities that formed part of the *UACES Research Network on European Energy Policy* which they initiated, and so competently and energetically led.

Disclosure statement

No potential conflict of interest was reported by the author.

Funding

This research has benefitted from the author's post-doctoral project 'Med eller utan EU? Konsekvensen av nationella utrikespolitiska val för Europeiska unionens effektivitet som global aktör' (2011–2014), funded by the Swedish Research Council (VR) Dnr 435-2011-1106, and the author's Research Fellowship in the Special Research Programme on International Studies at the Swedish Institute of International Affairs (UI, 2014-2018), financed by the Swedish Foreign Ministry.

ORCID

Anke Schmidt-Felzmann ⓘD http://orcid.org/0000-0003-4878-9576

References

Ahlborg, K., and K. Danielson. 2016. "Farligt för Sverige men kommunerna bestämmer." *Aftonbladet*, December 13.

Alvarez Sanz, B. 2017. "The Yukos Saga Reloaded: Further Developments In The Interplay Between Domestic Legislations And Provisionally Applied Treaties." *International Law And Politics* 49: 587–605.

Andersen, S. S., and N. Sitter. 2015. "Managing Heterogeneity in the EU: Using Gas Market Liberalisation to Explore the Changing Mechanisms of Intergovernmental Governance." *Journal of European Integration* 37 (3): 319–334. doi:10.1080/07036337.2014.953947.

Baltic News Network. 2017a. "Government Decides Not to Support Ventspils Involvement in Nord Stream 2." April 21.

Baltic News Network. 2017b. "BATL: Economic Aspects are Just as Important as Security in Nord Stream 2." May 2.

Baltic News Network. 2017c. "Programme: Nord Stream 2 Developers Pause with Ventspils' Involvement." May 2.

Baltic News Network. 2017d. "Latvian Government Asked to Compensate Profits Lost from Rejecting Nord Stream 2." June 21.

Bech-Bruun. 2017. "Høringssvar på vegne af Nord Stream 2 AG, 1-27." August 9.

Bergamt Stralsund. 2018. "Bergrechtliches Genehmigungsverfahren gemäß § 733 Abs. 7 Satz 1 Nr. 7 BBergG für die Errichtung und den Betrieb der Transit Rohrleitung „Nord Stream 2" Narva-Bucht (RUS) - Lubmin (DEU) im Bereich des deutschen Festlandsockels." March 16.

Bilgin, M. 2011. "Energy Security and Russia's Gas Strategy: The Symbiotic Relationship between the State and Firms." *Communist and Post-Communist Studies* 44: 119–127. doi:10.1016/j.postcomstud.2011.04.002.

Boussena, S., and C. Locatelli. 2017. "Gazprom and the Complexity of the EU Gas Market: A Strategy to Define." *Post-Communist Economies* 29 (4): 549–564. doi:10.1080/14631377.2017.1349667.

Bouzarovski, S., and M. Konieczny. 2010. "Landscapes of Paradox: Public Discourses and Policies in Poland's Relationship with the Nord Stream Pipeline." *Geopolitics* 15 (1): 1–22. doi:10.1080/14650040903420362.

Bundesministerium für Wirtschaft und Energie (BMWI). 2018. "Antwort auf Schriftliche Frage an die Bundesregierung im Monat Februar 2018." *Frage Nr. 74*, February 14.

Bundesministerium für Wirtschaft und Energie (BMWI). 2016. "Kleine Anfrage betr "Erweiterung der Ostseepipeline Nord Stream 2." *BT Drucksache 18/10127*, November 15.

Bundesnetzagentur. 2019. "Bundesnetzagentur Orders Immediate Implementation of OPAL Judgment of European Court." September 13.

Court of Justice of the European Union (CJEU). 2019. "Case T 883/16, Judgment of the General Court (First Chamber, Extended Composition)." September 10.

Cruz, T. (US Senator, Texas). 2019. "Final NDAA 2020 Includes Sens Cruz, Shaheen Bill to Stop Russia's Nord Stream 2 Pipeline." *Press Release 202-228-7561*, December 9.

Dralle, T. M. 2018. *Ownership Unbundling and Related Measures in the Energy Sector, Foundations, the Impact of WTO Law and Investment Protection*. Dresden: Springer.

Dudek, J., and A. Piebalgs. 2017. "'Nord Stream 2 and the EU Regulatory Framework: Challenges Ahead', Policy Brief Issue 2017/26."*Florence: Robert Schuman Centre for Advanced Studies*.

Edberg, K., A.-L. Fransson, and I. Elander 2017. "Island and the Pipeline: Gotland Facing the Geopolitical Power of Nord Stream." *Centre for Urban and Regional Research, Report 71*, Örebro University.

Eichberger, F. S. 2019. "Nord Stream 2: Arbitration Notices from Moscow." October 1. https:// voelkerrechtsblog.org/nord-stream-2-arbitration-notices-from-moscow/

Energistyrelsen. 2019a. "Resumé af nationale høringssvar." October 30.

Energistyrelsen. 2019b. "Tilladelse Til Nord Stream 2 Naturgasrørledningerne I Østersøen." October 31.

European Commission. 2014. "Breakthrough: 4,6 Billion Dollar Deal Secures Gas for Ukraine and EU." *IP/14/1238*.

European Commission. 2017a. "Commission Seeks a Mandate from Member States to Negotiate with Russia an Agreement on Nord Stream". *IP/17/1571*.

European Commission. 2017b. "Questions and Answers on the Commission Proposal to Amend the Gas Directive (2009/73/EC)." *MEMO/17/4422*.

European Commission. 2017c. "Recommendation for a Council Decision authorizing the opening of negotiations on an agreement between the European Union and the Russian Federation on the operation of the Nord Stream 2 pipeline", at http://www.politico.eu/wp-content/uploads/2017/07/NS-Draft-Mandate.pdf.

European Commission. 2018. "Commission Welcomes WTO Ruling Confirming Lawfulness of Core Principles of the EU Third Energy Package." *IP/18/4942*.

European Commission. 2019a. "Energy Union: Commission Welcomes Tonight's Provisional Political Agreement to Ensure that Pipelines with Third Countries Comply with EU Gas Rules." *IP/19/1069*.

European Commission. 2019b. "Reply to Nord Stream 2 Letter of 8 July." *DG Trade*, July 26.

European Parliament (EP). 2007. "The Nord Stream Gas Pipeline Project and Its Strategic Implications." http://www.europarl.europa.eu/thinktank/en/document.html?reference=IPOL-PETI_NT(2007)393274

European Parliament (EP). 2015. "Energy Union: Nord Stream and Involving EP - Statement from ITRE Chair Buzek." November 18.

European Parliament (EP). 2018. "MEPs Commend Ukraine's Reform Efforts and Denounce Russian Aggression." *AFET*, December 12.

European Parliament (EP). 2019. "Natural Gas: Parliament Extends EU Rules to Pipelines from non-EU Countries." April 4.

Finon, D., and C. Locatelli. 2008. "Russian and European Gas Interdependence: Could Contractual Trade Channel Geopolitics?" *Energy Policy* 36: 423–442. doi:10.1016/j.enpol.2007.08.038.

Folketing. 2017. "L43 Forslag til lov om ændring af lov om kontinentalsoklen."

Freeport of Ventspils. n.d. "Board." http://www.portofventspils.lv/en/port-administration/board/

German Foreign Ministry. 2017. "Foreign Minister Gabriel and Austrian Federal Chancellor Kern on the Imposition of Russia Sanctions by the US Senate." June 15.

Godzimirski, J. M. 2011. "Nord Stream: Globalization in the Pipeline?" In *Russia's Encounter with Globalization: Actors, Processes and Critical Moments*, edited by J. Wilhelmsen and E. W. Rowe, 159–184. London: Palgrave Macmillan.

Goldthau, A., and N. Sitter. 2014. "A Liberal Actor in A Realist World? the Commission and the External Dimension of the Single Market for Energy." *Journal of European Public Policy* 21: 1425–1472. doi:10.1080/13501763.2014.912251.

Goldthau, A., and N. Sitter. 2015. "Soft Power with a Hard Edge: EU Policy Tools and Energy Security." *Review of International Political Economy* 22 (5): 941–965. doi:10.1080/09692290.2015.1008547.

Government Offices of Sweden. 2016a. "Ansökan om tillstånd för gasledningen Nord Stream 2 har kommit till Regeringen." September 16.

Government Offices of Sweden. 2016b. "Informationsmöte den 13d ecember 2016 – Minnesanteckningar, made public by Swedish Public Radio." https://sverigesradio.se/diverse/appdata/isidor/files/94/93988121-7be4-46e7-95d2-d57c27af014b.pdf

Government Offices of Sweden. 2018. "Decision on the Application from Nord Stream 2 AG." *Ministry for Enterprise and Innovation*, June 7.

Handelsblatt. 2019. "Offensichtliche Trickserei": Nord Stream 2 Will EU-Regulierung Entkommen." November 7.

Henderson, J., and A. Moe. 2016. "Gazprom's LNG Offensive: A Demonstration of Monopoly Strength or Impetus for Russian Gas Sector Reform?" *Post-Communist Economies* 28 (3): 281–299. doi:10.1080/14631377.2016.1203206.

Herbert Smith Freehills. 2019. "Summary of NS2 Legal Concerns, 14 June, Made Public by Euractiv." https://www.euractiv.com/wp-content/uploads/sites/2/2019/07/Legal-concerns-Nord-Stream.pdf

Herranz-Surralles, A., I. I. Solorio Sandoval, and J. Fairbrass. 2020. "Renegotiating Authority in The Energy Union." *Shifting Patterns Of Integration in 'Hybrid Areas'a Framework for Analysis." Journal of European Integration* 421: 18. doi: 10.1080/07036337.2019.1708343.

Högselius, P. 2013. *Red Gas*. New York: Palgrave Macmillan.

Ifflander, H., and L. Soneryd. 2014. "The Relative Power of Environmental Assessment as a Governance Tool: Organization and the Case of the Nord Stream Gas Pipeline." *Impact Assessment and Project Appraisal* 32 (2): 98–107. doi:10.1080/14615517.2014.896083.

Karm, E. 2008. "Environment and Energy: The Baltic Sea Gas Pipeline." *Journal of Baltic Studies* 39 (2): 99–121. doi:10.1080/01629770802031200.

Koivurova, T., and I. Pölönen. 2010. "Transboundary Environmental Impact Assessment in the Case of the Baltic Sea Gas Pipeline." *International Journal of Marine and Coastal Law* 25: 151–181. doi:10.1163/157180910X12665776638588.

Langlet, D. 2014a. "Nord Stream, the Environment and the Law: Disentangling a Multijurisdictional Energy Project." *Scandinavian Law* 59 (2): 179–205.

Langlet, D. 2014b. "Transboundary Transit Pipelines: Reflections on the Balancing of Rights and Interests in Light of the North Stream Project." *International & Comparative Law Quarterly* 63 (4): 977–995. doi:10.1017/S0020589314000232.

Lee, R., T. Egede, L. Frater, S. Vaughan. 2008. "Legal Implications of the Nord Stream Project, European Parliament's Committee on Foreign Affairs." *EP/EXPO/B/AFET/2008/02*

Lidskog, R., and I. Elander. 2012. "Sweden and the Baltic Sea Pipeline: Between Ecology and Economy." *Marine Policy* 36: 333–338. doi:10.1016/j.marpol.2011.06.006.

Lidskog, R., Y. Uggla, L. Soneryd. 2011. "Making Transboundary Risks Governable: Reducing Complexity, Constructing Spatial Identity, and Ascribing Capabilities." *Ambio* 40 (2): 111–120. doi:10.1007/s13280-010-0123-3.

Locatelli, C. 2014. "The Russian Gas Industry: Challenges to the 'gazprom Model'?" *Post-Communist Economies* 26 (1): 53–66. doi:10.1080/14631377.2014.874232.

Marhold, A. 2012. "In Too Deep – Russia, the Energy Charter Treaty and the Nord Stream Gas Pipeline." *Baltic Yearbook of International Law* 12: 303–315. doi:10.1163/22115897-90000097.

Mayer, S. 2008. "Path Dependence and Commission Activism in the Evolution of the European Union's External Energy Policy." *Journal of International Relations and Development* 11 (3): 251–278. doi:10.1057/jird.2008.12.

McGowan, F. 2011. "Putting Energy Insecurity into Historical Context: European Responses to the Energy Crises of the 1970s and 2000s." *Geopolitics* 16 (3): 486–511. doi:10.1080/14650045.2011.520857.

NDR. 2018a. "NABU Klagt Gegen Nord Stream 2." March 2.

NDR. 2018b. "Nord Stream 2 Stößt in Lubmin Auf Wohlwollen." October 10.

NDR. 2018c. "NABU Will Baubeginn Von Nord Stream 2 Verhindern." May 11.

NDR. 2018d. "Bau Der Erdgas-Pipeline Nord Stream 2 Gestartet." May 15.

NDR. 2018e. "Fett-Klumpen: Nord Stream Lässt Arbeiten Ruhen." May 30.

NDR. 2018f. "Kein Pipeline-Baustopp Gericht in Der Kritik." June 2.

NDR. 2018g. "Nord Stream 2: NABU Reicht Verfassungsbeschwerde Ein." July 3.

NDR1, Radio MV. 2018. "Schwesig Zu Nord Stream 2: Wir Entscheiden Selbst." May 17.

Nord Stream 2. 2017. "Nord Stream 2 Welcomes Decision regarding Pipe Storage at Port of Karlshamn." January 31.

Nord Stream 2. 2019a. "Letter to European Commission President Juncker Concerning the Directive Amending Directive 2009/73/EC." 12 April.

Nord Stream 2. 2019b. "Letter Addressed to the European Commission DG Trade, Directorate F2." July 8.

Nord Stream 2. 2019c. "Nord Stream 2 Calls on Court of Justice of the European Union to Annul Discriminatory Measures." July 25.

President of Russia. 2015. "Vladimir Putin Met with Vice-Chancellor and Minister of Economic Affairs and Energy of Germany Sigmar Gabriel." October 28. http://en.kremlin.ru/events/president/news/50582

Riley, A. 2010. "The EU-Russia Energy Relationship: Will the Yukos Decision Trigger a Fundamental Reassessment in Moscow?" *International Energy Law Review* 2: 36–40.

Riley, A. 2016. "Nord Stream 2: A Legal and Policy Analysis", *CEPS Special Report no. 151.*

Riley, A. 2018. "A Pipeline Too Far? EU Law Obstacles to Nord Stream 2." *International Energy Law Review*, March.

Romanova, T. 2016. "Is Russian Energy Policy Towards the EU Only about Geopolitics? The Case of the Third Liberalisation Package." *Geopolitics* 21 (4): 857–879. doi:10.1080/14650045.2016.1155049.

Roth, M. 2011. "Poland as a Policy Entrepreneur in European External Energy Policy: Towards Greater Energy Solidarity Vis-à-vis Russia?" *Geopolitics* 16 (3): 600–625. doi:10.1080/14650045.2011.520865.

Sayuli, F., I. Elander, R. Lidskog. 2011. "Framing Issues and Forming Opinions: The Baltic Sea Pipeline in the Swedish Media." *European Spatial Research and Policy* 18 (2): 95–110. doi:10.2478/v10105-011-0015-y.

Schill, S. 2017. "Memorandum on Draft Bill to Amend the Danish Continental Shelf Act, L43, Bilag 3." *Folketing.*

Schmidt-Felzmann, A. 2019a. "The Commercial Deals Connected with Gazprom's Nord Stream 2 A Review of Strings and Benefits Attached to the Controversial Russian Pipelines." *Think Visegrad – V4 Think Tank Platform*, March 2019, https://think.visegradfund.org/wp-content/uploads/Felzmann_The-commercial-deals-of-Gazproms-Nord-Stream-2.pdf

Schmidt-Felzmann, A. 2019b. "Between Geopolitics and Market Rules: The EU's Energy Interdependence with Russia." In *Post-Crimea Shift in EU-Russia Relations: From Fostering Interdependence to Managing Vulnerabilities*, edited by K. Raik and A. Racz, 142–161. Tallinn: ICDS/RKK.

Schmidt-Felzmann, A. 2019c. "Negotiating at Cross Purposes: Conflicts and Continuity in the EU's Trade and Energy Relations with Russia, Pre- and Post-2014." *Journal of European Public Policy* 26 (12): 1900–1916. doi:10.1080/13501763.2019.1678057.

Sharples, J. D. 2016. "The Shifting Geopolitics of Russia's Natural Gas Exports and Their Impact on EU-Russia Gas Relations." *Geopolitics* 21 (4): 880–892. doi:10.1080/14650045.2016.1148690.

Sjöfartsverket. 2019. "*Underrättelser För Sjöfarande* (UFS) - a Weekly Newsletter with Updates on Changes to Maps and Ongoing Disruptions Concerning Swedish Waters." www.sjofartsverket.se

Stoddard, E. 2017. "Tough Times, Shifting Roles: Examining the EU's Commercial Diplomacy in Foreign Energy Markets." *Journal of European Public Policy* 24 (7): 1048–1068. doi:10.1080/13501763.2016.1170190.

SVT. 2016. "Wallström: Vi kan inte ensamma stoppa Nord Stream." September 8.

SVT. 2017. "Regeringen: Vi har inte ändrat oss om Nord Stream 2." January 30.

Talus, K. 2017. "Application of EU Energy and Certain National Laws of Baltic Sea Countries to the Nord Stream 2 Pipeline Project." *Journal of World Energy Law and Business* 10: 30–42.

Talus, K., and M. Wüstenberg. 2017. "Risks of Expanding the Geographical Scope of EU Energy Law." *European Energy and Environmental Law Review* 26 (5): 138–147.

TV2/Bornholm. 2017. "Fakta om Nord Stream 2." January 27.

United Nations Convention on the Law of the Sea (UNCLOS). 1982. *Adopted* December, 10, in force since November, 16 1994." 1833 UNTS 397.

United States Code. 2019. "Title 22 Imposition of Sanctions with respect to Persons Engaging in Transactions with the Intelligence or Defense Sectors of the Government of the Russian Federation, 22 USC 9525." https://uscode.house.gov/view.xhtml?req=(title:22%20section:9525%20edition:prelim)

UOKiK. 2016. "UOKiK Issues Objections to a Concentration - Nord Stream 2." July 22.

UOKiK. 2018. "UOKiK against Nord Stream 2." May 9.

US Congress. 2017a. "S.1221 - Countering Russian Influence in Europe and Eurasia Act of 2017.", June 6.

US Congress. 2017b. "H.R.3364 - Countering America's Adversaries through Sanctions Act (Now Public Law No: 115–44), August 2.

US Congress. 2018a. "H.R.6384 - Countering Russian Power Plays Act." July 16.

US Congress. 2018b. "H.Res.1035 - Expressing Opposition to the Completion of Nord Stream II, and for Other Purposes." July 26.

US Congress. 2019a. "H.Res.116/S.Res.27 - A Resolution Calling for A Prompt Multinational Freedom of Navigation Operation in the Black Sea and Urging the Cancellation of the Nord Stream 2 Pipeline." February 8/January 24.

US Congress. 2019b. "H.R.1081 - Protect European Energy Security Act." February 7.

US Congress. 2019c. "H.R.2023 - Protect European Energy Security Act." May 22.

US Congress. 2019d. "H.R.4818/S.1441 - Protecting Europe's Energy Security Act of 2019." May 14.

US Congress. 2019e. "H.R.3206 - Protecting Europe's Energy Security Act of 2019." June 11.

US Congress. 2019f. "S.1830 - Energy Security Cooperation with Allied Partners in Europe Act of 2019." June 13.

US Congress. 2019g. "H.R.2500 - National Defense Authorization Act (NDAA) for Fiscal Year 2020." February 5, in force since 21 December 2019.

US Department of State. 2019. "Public Designations Due to Significant Corruption of Latvian and Cambodian Officials." December 10. https://www.state.gov/public-designations-due-to-significant-corruption-of-latvian-and-cambodian-officials/

Vihma, A., and M. Wigell. 2016. "Russia's Geoeconomics and the Nord Stream II Pipeline." *Global Affairs* 2 (4): 377–388. doi:10.1080/23340460.2016.1251073.

Voon, T., and A. Mitchell. 2017. "Ending International Investment Agreements: Russia's Withdrawal from Participation in the Energy Charter Treaty." *American Journal of International Law* 111: 461–466.

Wojcieszak, Ł. 2017. "Nord Stream 2 Pipeline. Role of German-Russian Energy Cooperation for Poland." *American Journal of Sociological Research* 7 (3): 85–89.

EU foreign policy and energy strategy: bounded contestation

Richard Youngs

ABSTRACT

The external dimensions of energy security are subject to distinctive forms of contestation within the European Union (EU). Contestation related to the external dimensions of energy strategy is significant, but bounded. EU institutions and member states have contained contestation in at least some areas of external energy strategy, through both de-politicization dynamics and new mechanisms of horizontal coordination. The article demonstrates and explains such bounded contestation in three areas of EU international policies: security of supply, climate security and wider security challenges. It explains the different logics that underpin the management of divergence, and relates these to analytical debates over EU foreign and security policy. The article stresses the importance of understanding how contestation over energy policy is nested within a broader set of geopolitical imperatives.

I. Introduction

This article focuses on the relationship between EU foreign policy and energy strategies. It examines how debates over foreign and security policies influence the degree of internal EU contestation over energy and climate policy. While other contributions on EU external energy policy in this volume look at how energy considerations spill-over to the foreign-policy sphere, the article tackles this linkage in the reverse direction. The foreign-policy impact on energy strategy is important and has become increasingly controversial. The EU claimed that the Energy Union was in part about strengthening the geopolitical aspects of energy policy. When the Energy Union was launched, analysts debated about whether this geopolitical focus would intensify tensions between the EU's energy and broader foreign policies – or in terms of this volume's main conceptual focus, whether geopolitical issues risked intensifying contestation over energy policies (Buchan and Keay 2015; Andoura and Vinois 2015; Scholten et al. 2015; Helm 2015; Fischer and Geden 2015).

This article argues that in practice the link between foreign and energy policy has turned out to be complex and varied. The foreign affairs aspects of energy policy provide an example of how substance-based contestation can be important even where the kind of formal disputes over legal authority covered throughout this volume are not prominent. However, they also shed light on the factors *limiting* such contestation. The external dimensions of energy policy contain much of that is about contestation, but also much of that points to the *mitigation* of internal EU differences. The article's focus on the way in which energy-policy contestation is tempered is categorically *not* meant to deny or downplay the serious policy differences that remain between member states and EU institutions; rather, it is more modestly to point out that alongside the well-known tensions over energy policies there are also at least some factors working in the opposite direction of better-managed contestation.

In short, the article shows that foreign and security-policy concerns can intensify internal EU contestation over the classical elements of energy policy, but they can also dampen it. The article suggests a notion of *bounded contestation* to capture the impact of the foreign-policy sphere on EU energy policy – and it shows how this concept relates to academic perspectives on EU foreign and security policy (section 2). The article unpacks how this bounded contestation has taken root across three aspects of foreign and security policy: security of supply (section 3); climate security (section 4); and wider security strategies (section 5). These are chosen both because they are the areas of policy where bounded contestation is most evident, but also because contestation is limited through a different means or logic in each case. The article concludes (section 6) by tying these three areas back to the concept of bounded contestation and current, overarching academic debates about EU external relations.

II. EU foreign policy: towards bounded contestation?

The introduction to this volume maps out an analytical framework for assessing how much contestation exists over EU energy policy; whether this is sovereignty-based or substance-based contestation; and how such contestation is contained (Herranz-Surralles, Solorio and Fairbrass, 2020). This article examines how EU foreign and security policy relates to this contestation over European energy policies. In order to shed light on this link, it is instructive to draw from recent theoretical debates about EU foreign and security policy. These debates add quite specific elements to the volume's core analytical concepts of contestation and renegotiation of EU decision-making authority.

The management of divergence and contestation has long been a central pillar of analytical work on EU foreign and security policy. It is well known that EU foreign and security policy remains largely, although not purely, inter-governmental in nature. National governments remain primary actors, especially on the harder dimensions of international security. The dynamics of authority are complicated where foreign and security-policy aims require the deployment of policy instruments that are either supra-nationalized or at least more Europeanized than traditional diplomacy – these include trade, migration, development aid and energy policies. Of specific relevance to this volume, a gap in institutional dynamics has opened up between the development of supranational competences over internal energy and climate issues, on the one hand, and the still largely intergovernmental nature of EU foreign and security policy, on the other hand (Birchfield and Duffield 2011).

Over many years, foreign-policy analysts have debated what these formal divisions of competence mean for the de facto coherence and efficiency of EU decision-making. Two broad strands of analysis can be identified in this sense – each of which has evolved in recent years in ways that are significant for the question of policy contestation within the union.

One analytical strand argues that contestation over foreign and security policy is significant, not well managed by EU institutional structures and has if anything become more acute in the last several years. The standard realist critic is that the mechanisms of EU foreign policy-making rarely succeed in unifying member-state positions (Hyde-Price 2007). Some insist that changes to the international system have pulled European governments in increasingly different directions, making the EU a less unique and united or self-enclosed foreign-policy system (Rosato 2011). More recently, there has been much talk of the 're-nationalization' of foreign policies – the new High Representative, Josep Borrell has suggested this is now a powerful dynamic that undercuts effective policy coordination (Borrell 2019). Much recent writing is about the return of standard national-interest realpolitik and how this has fragmented the internal dynamics of EU foreign and security policy (Biscop 2015).

Analysts have suggested that the foreign-policy sphere is one where the dynamics of 'new intergovernmentalism' have been especially evident, especially in the fact that member states set up the European External Action Service (EEAS) outside the Commission as a de novo body in order to recapture control of the foreign-policy agenda (Bickerton, Hodson, and Puetter 2015; in response, Schimmelfennig 2015; Amadio Vicere 2016). Other experts stress that European foreign policy has become more multi-layered, with the national level of external agency actually gaining influence and the dynamics of classical geopolitical balancing returning to intra-European relations (Wivel and Wæver 2018; Hill, Smith, and Vanhoonacker 2017). In some instances, widening differences have been dealt with by some member states forming small breakaway groupings to pursue certain policies outside the common EU framework (Aggestam and Bicchi, 2019). There is also emergent academic debate about a deepening 'politicization' creeping into the sphere of EU foreign and security policy, opening up divergences and areas of contestation even beyond the longstanding differences between member states' interest-based calculations (Costa 2018).

A contrasting strand of analytical work argues that the EU's foreign and security policy has gradually taken on a more common identity and found ways of closing the differences between member states and EU institutions. Analysts commonly stress how internal processes of EU coordination have gradually generated a dynamic of socialization around certain cooperative security norms. They hold that embedded institutional dynamics have bred common outlooks between member states around rule-based international action (Thomas 2011; Keukelaire and Delreux 2014; Juncos and Pomorska 2015). Despite formal institutional fragmentation, cooperative reflexes and horizontal coordination may often limit contestation – albeit to varying degrees across different policy domains. Sociological approaches to governance across borders have resonance in the foreign-policy sphere, where EU-level authority is less formal than earned de facto through repeat policy cycles and years of regularized policy coordination in which EU agents prove themselves valuable to member states. The result is that in at least some areas of foreign policy, analysts point to the EU's success in mitigating divergences between member states and EU institutions – with common rules and standards helping a process of de-politicization (Börzel and van Hüllen 2015; Gstöhl and Schunz 2016).

This convergence is often moulded around norm-based notions of security and the external projection of EU functional rules as core, common tools of foreign and security policy (Telo, 2007; Manners 2008; Birchfield 2013; Lavenex 2014). Far from EU foreign policy losing momentum and unity in recent years, some point to a deepening trend towards 'collective securitization' as member states have come commonly to define external strategic challenges in a way that draws national and European level strategic approaches together (Sperling and Webber 2019). Many analysts have in recent years argued that policy-making dynamics have taken root that have member states and different parts of the European Commission coming together around common understandings of security – and that these dynamics can and do transcend the patterns of formal authority-conferral between these different actors (*Ibid.*). Some analysts have suggested nuance to the new intergovernmentalism, namely that the EEAS has gradually developed some autonomy as an agenda-setter beyond the control of member states and that this has brought more far-reaching convergence than that achieved through intergovernmental bargaining alone (Morillas 2019).

In addition to the dynamics of internal coordination, an additional analytical perspective is that external factors have also increasingly encouraged member states and EU institutions to dilute mutual contestation. A common line of reasoning is that the EU's relative decline in global power has acted as a catalyst for more effective foreign-policy cooperation, in particular after the post-2009 economic crisis so clearly left the union in a weaker position internationally (Wright 2012). Modifying the standard structural realist perspective, it is argued that precisely because the international order is changing shape and strategic imperatives becoming more challenging, a new dynamic has come to propel deeper EU foreign-policy unity (Krotz and Maher 2012). Interestingly, theorists have sought to modify institutionalist approaches to account for these kinds of common policy changes in response to external power-shifts (Schmidt 2010). Some experts see pressures towards deeper strategic cooperation resulting from exogenous forces representing an even deeper threat to Europeans' basic ontological security, beyond traditional notions of security (Lucarelli 2019). Notably, this ethos is prominent in the 2016 EU Global Strategy, which claims that common EU security policies are now taking shape in large measure because the external environment threatens the whole European project (European Union 2016).

The crucial question then arises of what implications these theoretical debates on EU foreign and security policy have for this volume's analytical framework of contestation in energy policy. The link is normally considered the other way round, with a focus on how differences over energy policy spill over into differences over foreign policy – especially in terms of market and geopolitical approaches to energy having contrasting implications for EU foreign policies (Goldthau and Sitter 2015a; Van der Linde 2007). Yet the article will show why there are important reasons also to consider how wider foreign-policy debates feed into and condition EU energy and climate positions.

We can hypothesize that the degree of contestation within EU foreign and security-policy debates will condition the degree of contestation in energy policy. The first strand of academic enquiry on EU foreign policy outlined above is varied in its theoretical foundations and practical applications, yet in general resonates with the notion that contestation is increasing in intensity and becoming a powerful dynamic within foreign and security policy. In line with this, and in linking foreign-policy debates to this volume's analytical framework, a first scenario is that as contestation deepens in core foreign and security policy, so we would expect a spill-over increase in contestation in the external aspects of energy policy.

However, there is a second scenario more in line with analytical work that points to the attenuation of contestation in foreign and security policy: the management and narrowing of differences within core EU foreign and security-policy issues can be expected to assist in the management of contestation within energy policy. In accordance with the different theoretical perspectives within foreign-policy debates, these 'limiting' factors could derive from internal coordination mechanisms within foreign-policy decision-making and/or from the impact of substantive challenges that prompt deeper EU coordination and de-politicize energy policy at least to some degree.

If this latter strand of theoretical work has validity, we might expect to see foreign-policy concerns helping to limit the degree of internal contestation over energy policy. A notion of what could be labelled *bounded contestation* would encapsulate this. This can be defined as a situation where differences exist between different institutional actors, but kept within limits that allow for some degree of common policy positions to be developed. Bounded contestation would be identified when initial divergences in preferences over energy policy are narrowed, managed and de-politicized as and when energy issues are nested within a wider set of concerns about foreign and security-policy interests. Differences between member states and EU institutions would remain, but of a second-order rather than first-order type: that is, not as principled differences over core objectives, but rather differences of judgement over tactical approaches to achieving or implementing those goals.

The general claim being suggested here is that links *between* different policy spheres – in this case from foreign policy to energy policy – can play a role in managing divergence and generating a dynamic of de-politicization. The claim is categorically *not* that contestation does not exist or is not serious; rather, it is that there are dynamics that work towards contestation being less all-consuming than it would otherwise be. The argument should not be taken as dismissing politicization, but as suggesting that politicization can co-exist with at least partially off-setting dynamics of de-politicization.

The article is concerned with testing this scenario of bounded contestation. It shows three senses in which bounded contestation has taken shape in the link between general foreign and security policies, on the one hand, and energy policies, on the other hand. In the terms set by the introductory framework, each case has seen a reconfigured conferral of authority not in formal legal terms but in terms of the de facto manner in which patterns of EU policy-making authority have shifted in response to energy-policy challenges. The three cases are: security of supply; climate security; and wider security strategies These three domains are chosen because they each show a slightly *different logic* through which contestation is either managed or limited. In each case, material is drawn from EU and member-state strategy documents, as well as from several dozen anonymous interviews with EU policy-makers and off-the-record roundtables organized with diplomats between September 2013 and September 2019.

III. Security of supply: inter-issue balances and horizontal coordination

The first area of bounded contestation lies in relation to security-of-supply strategies. The logic through which contestation has been managed here involves inter-issue balances in conjunction with new horizontal coordination mechanisms – while external energy-market changes have also helped as a flanking factor.

In the 2010s, the EU has signed an increased number of external security-of-supply agreements. As of 2019, the EU has 18 energy dialogues with individual countries, 29 memorandums of understanding and declarations on energy cooperation, and 7 regional initiatives (covering the Eastern Partnership, Union for the Mediterranean, Africa, the Gulf Cooperation Council, ASEAN and the Caspian area, along with the Energy Community Treaty) (European Commission 2015).

As the number of these accords has multiplied, the EU has set up improved coordination mechanisms that have given the EU's foreign-policy machinery a more systematic and significant role in external energy agreements. In July 2015 the EEAS launched its Energy Diplomacy concept, which stated that 'foreign policy instruments and channels for the engagement should be used to open up opportunities for cooperation with increasingly important producing and transit countries' (Council of the EU 2015). The initiative was aimed at getting the EEAS and member-state foreign-policy establishments engaged in energy dialogues and providing high-political, diplomatic influence over what have in the past been rather low-profile technical, Commission-led initiatives. Horizontal coordination has also taken place through a Strategy Group on International Energy that since 2012 brings together energy-related Commission DGs with diplomats from the EEAS. The EEAS has also brought in a team of energy specialists in order to help improve coherence between security and energy strategies.

Of course, differences persist over the security of supply questions. Some member states and EU institutions have expressed doubts over some new pipeline projects. They are sceptical about the Southern Gas Corridor on the grounds that this risks proving uneconomic, is unlikely to bring significant quantities of gas into most parts of the EU, flouts climate goals and may now even be set to carry Russian gas – largely negating its supposed geopolitical rationale (Siddi 2019). Unsurprisingly, DG Clima, leading on climate action, tends to fear DG Energy pushing the overall policy-balance too far away from climate goals towards investment in new pipelines and supply partnerships (Far and Youngs 2015).

While there have been serious tensions over security-of-supply questions, however, the EU has sought to develop an overall approach that balances different perspectives. Member states, the European Commission and the EEAS agreed on the general need for the Energy Diplomacy initiative, when in previous years similar proposals did not gain sufficient consensus to advance in practice (Herranz-Surrallés 2016). While the Commission has been happy to play a lead role in security-of-supply projects, the EU has not put huge amounts of grant money into the Southern Gas Corridor – to date negligible 4 million euros from the Connecting Europe Facility. The EU has refrained from formal support for Turkish Stream, the first segment of which was opened in November 2018; even though this will bring supplies into European markets through Turkey, such capacity is unlikely to be essential. While Germany and some other states remain uneasy about attempts to cut Russia out of supply routes into Europe, they have accepted the basic case for an EU role in energy-supply diversification and a balance of approaches towards supply issues (Goldthau and Sitter 2015b). These factors and balances have helped reduce the level of contestation, without ensuring complete unity between member states or EU institutions (De Jong 2016).

Despite the contestation over Nordstream 2 – covered in detail elsewhere in this volume (Goldthau and Sitter 2020; Schmidt-Felzmann 2020), and the external energy issue that attracts by far the most analytical attention – member states still formally support the Energy Union's original aspiration to ensure that one member state should not be able to decide on a pipeline project or energy deal that undermines the security of other member states. Again, a degree of compromise is evident on this question. Most member states rejected proposals to give the Commission new powers to negotiate gas contracts on behalf of the whole European Union. Instead, the Commission has gained more modest powers to assess the compatibility of member states' bilateral energy agreements with EU rules. A 2012 EU decision on an Information Exchange Mechanism on Intergovernmental Agreements on energy (Decision 994/2012) encourages member states to share information on their deals with third countries; new rules in 2017 require governments to share information on intergovernmental agreements for gas before they are signed rather than only after they are concluded (Decision 684/2017). The Commission is then tasked with offering advice on an agreement's compatibility with EU energy goals. A balance has hence been struck. European governments have not been willing to give the Commission powers to vet energy contracts concluded by member states with third countries, but have sought a framework for managing the most serious of disputes over particular energy contracts (European Parliament 2017).

Overarching supply trends have served to underpin an approach that balances support for pipelines with market-based norms and de-carbonization aims. The EU is likely to require relatively modest increases in natural gas supplies over the medium term (BP 2018). Deeper internal coordination through the Gas Market Directive is already helping to reduce external vulnerabilities. Security of supply has lost some of its do-or-die visceral importance within internal EU deliberations. It is far from the dominant and divisive axis of EU energy policy as was the case in the early 2010s, when the union's external energy vulnerabilities appeared especially pressing (Szulecki and Westphal 2014). The Commission has pursued a combination of both energy transition funding around the world and new incoming pipeline projects.

Indeed, a crucial aspect of the balanced approach is that the investment of resources in de-carbonization has increased both inside the EU and in the union's external funding. Alongside the raft of new strategic partnerships and other external energy-supply initiatives, the EU support for third countries' energy transitions and climate adaptation has become the fastest rising area of external funding (European Commission 2017). The EU's climate financing has grown dramatically and totalled 20.7 billion euros in 2017, including funds from the European Commission, member states and the European Investment Bank (European Commission 2019). In March 2019, the European Parliament insisted that external climate funding be significantly increased under the Multi-Annual Financial Framework for 2020–2027, and in particular account for a minimum 45% of funds allocated under the new umbrella Neighbourhood, Development and International Cooperation Instrument (New Europe 2019). In a more specific example, in October 2019 the EU announced a 100 million euro climate financing initiative in Brazil – significant against the backdrop of acute political tensions generated over the then on-going Amazon fires (New Europe 2019b). The Commission has also begun to include tougher climate-related conditions in its trade agreements (Climate Action 2018).

Those working on climate policy have gained influence and resources within the EU's foreign policy, somewhat blunting their criticism of the union's security-of-supply agreements and support for pipeline capacity. Commissioner Miguel Arias Cañete said in early 2018: 'energy transition … remains our answer to the geopolitical uncertainties we are facing' (Arias Cañete 2018). The Commission's *A clean planet for all* strategy document, published at the end of 2018, claims that if the EU meets all its targets and commitments related to internal energy transition, its dependence on external energy supplies will fall from 55% in 2018 to 20% in 2050 (European Commission 2018, 8). And the EU has also made linkages in the reverse direction, with renewables projects themselves feeding into security-of-supply policies: the Union has become more mercantile, insisting that many EU-supported renewables projects in third countries be used to increase energy supplies into Europe – the large Desertec project failed in large measure because the Moroccan government resisted it for this reason (Escribano 2019).

IV. Climate security: shared concerns?

The second area of bounded contestation is found in climate-security policy. Here the logic of managing contestation comes from a degree of substantive convergence of security understandings. An increased and shared concern over the security-policy impli-cations of climate change has helped bring the EU's security and climate-policy commu-nities somewhat into line with each other, where previously these were commonly engaged in stronger contestation.

The EU has become a lead player in climate security. The EU was one of the first organizations to identify climate change as a security issue – as a 'threat multiplier.' The Union has gradually put in place a collection of policy initiatives designed to mainstream climate-related factors within its foreign and security policies (Youngs 2014). Climate security is given prominence in the 2016 Global Strategy (European Union 2016). In February 2018 Council Conclusions committed the EU to doing more on all aspects of climate security, promising to 'further mainstream the nexus between climate change and security in policy dialogue, conflict prevention, development and humanitarian action and disaster risk strategies' (Council of the EU 2018). In June 2018, on the tenth anniver-sary of the first 'threat multiplier' paper in 2008, the EU promised more of a security-led role in climate issues, in particular through the High Representative beginning to play more of a role in this area. There was an internal debate over the creation of new working groups on climate-security risks, a watch centre, more early warning and new discussions on climate within various levels of foreign policy-making (Fetzek and van Schaik 2018). The Council issued a further statement in February 2019 reiterating a commitment not only to reducing emissions and engaging in the COPs process, but to tackling climate change as an 'existential' issue of international security (Council of the EU 2019). The Commission's *European Green Deal* of December 2019 reinforced this security linkage and called for the EU to craft a network of 'green alliances' as the backbone of its global geo-strategy (European Commission 2019b).

A relatively robust consensus now exists over the importance of this climate-security agenda. In general, there is a measure of agreement among the main EU actors and member states that climate change needs to be tackled as an issue that presents serious security risks to Europe as a whole. It has become part of the EU's internal dynamics of 'collective securitization' (Sperling et al. 2019). There are of course differences between institutions and governments on these questions. Diplomats within the EU foreign and security-policy community fear climate technocrats gaining too much weight and dis-torting necessary geopolitical calculation. Those working on climate and energy transi-tions tend to think of security as generic resilience of the EU energy system; those concerned with foreign policy tend to argue that the EU needs a more geopolitical and power-related understanding of security, including energy questions (Youngs 2014).

However, there is little really acute or hostile contestation over the importance of developing climate-security policies. In general terms, the intellectual case for focusing on climate issues as part of the security strategy is widely accepted across the EU institutions. An initially sceptical foreign and defence policy community has gradually adapted to the climate agenda. Conversely, climate-policy officers in the Commission have gradually moved beyond an initial ambivalence over the security community's involvement in climate issues. The Commission and member states have come to accept the need for a reshaped set of energy partnerships, looking beyond traditional hydrocarbon suppliers to the African and Asian countries housing materials like cobalt and lithium that will be key sources of power in the post-carbon era (Schunz 2012). The 2019 implementation report of the Global Strategy insists that the 'climate-security nexus' is the area where a 'joined-up' approach has been most successfully developed between different polices and parts of the EU, around the notion of 'sustainable security' (European External Action Service 2019, 28 and 40).

Crucially, a degree of convergence over climate security is underpinned by an approach that has avoided over-securitizing this agenda. Climate specialists feared that defence establishments would disingenuously ride the coattails of the climate-security agenda to gain more resources for themselves (Youngs 2014, chapter 5). Yet in practice, this agenda has mainly mobilized increases in development-oriented climate adaptation resources in third countries. The role of EU conflict stabilization policies or Common Security and Defence Policy (CSDP) missions in climate security remains limited by comparison. The EU is an undoubted leader in climate financing but has eschewed a really strong security-oriented approach to climate-related instability (Bergamaschi and Sartori 2018, 8).

These features of EU climate-security policies have helped limit contestation. Indeed, rather than a climate framing leading to divisive hard-security initiatives, mainstream climate policies have benefitted from a security framing. The Commission and other climate-change actors have deployed the climate-security agenda successfully as a means to help advance EU aims on climate-change diplomacy and the environment more broadly. The 2018 council conclusions express a broad agreement that the EU needs to change its security policy in light of climate change, but mostly stress the priority of the EU meeting its emissions-reduction commitments under the Paris accord (Council of the EU 2018). These dynamics have dampened the potential pushback against the climate-security agenda from within the EU institutions that a few years ago seemed likely to be of significant magnitude.

V. Wider security strategies: geopolitical constraints and de-politicisation

The third logic of bounded contestation comes from the impact of geopolitical contextual factors. More challenging security imperatives have acted to discourage member states from letting differences over energy policy reach a degree where they undermine the EU's geostrategic efficacy.

This has been seen in the case of Russia. As two other contributions to the volume cover Russia and the Nordstream 2 saga (Goldthau and Sitter 2020; Schmidt-Felzmann

2020), I do not delve into this issue in detail here. One different angle on this question is specific to this article, however: the seriousness of other, broader security challenges related to Russian actions in recent years. The EU has focused increasingly on wider geo-strategy beyond debates over energy supplies. As is well known, in relation to Russia's annexation of Crimea the EU has imposed sanctions on Russia, including a ban on the financing and export of innovative technologies to Russian oil companies. The deterioration in EU–Russia relations is well chronicled in the recent academic literature (Delcour and Kostanyan 2014; Nixey 2014; Sakwa 2015; Haukkala 2015; Raik 2016; Nitoiu 2016). In the context of contestation over Nordstream 2, it is significant to note that Germany has been one of the most vocal advocates of maintaining sanctions against Russia and is the member state that has pumped most money into Ukraine to help build its economic and institutional resilience against Russia (Youngs 2017).

The Ukraine conflict has encouraged the EU to begin mapping out a more geopolitical approach towards Russia. This approach is based on a mix of pressure and engagement. Notwithstanding self-evident differences between member states on strategic prefer-ences, a core EU line has gradually taken shape: the relatively far-reaching sanctions against Russia; increased cooperation through NATO; more active diplomatic engage-ment in the east of Ukraine, especially in respect of a mediation role in the Donbas conflict; enhanced support for Ukraine, Georgia and Moldova under the rubric of the Eastern Partnership; all balanced by a desire to explore other areas of strategic partnership with Russia in addressing other international concerns (Youngs 2017).

The standard line is that member states' different positions on energy policy have cut across broader strategic unity towards Russia. Yet this equation can also be assessed in the inverse direction: the most important EU priority in the last several years has been to maintain a meaningfully robust unity towards Russia in order to push back against threats to the European security order, and it is important to understand energy debates as being nested within this wider geopolitical context.

This clearly has not prevented high-level contestation between member states and EU institutions over Nordstream 2. It does, however, shed light on the efforts that have been made to manage this contestation. In February 2019, the Commission, member states and the European Parliament agreed on a compromise, under which Nordstream 2 could proceed but subject to tougher transparency standards and shared usage rules (Toplensky 2019). Even the hardest-line states on Russia, like the Netherlands, insisted this was a good compromise that would increase the influence of EU norms over Russia (Rutte 2019). To mitigate internal EU opposition to Nordstream 2, the German government insisted that Russia sign a new energy agreement with Ukraine as a quid pro quo for the new pipeline – arguing that in this way the project would not undercut the union's broader foreign policy aims in the region. The fourth round of talks between the EU, Russia and Ukraine collapsed in November 2019, leaving the future situation uncertain (New Europe 2019c). Moreover, a growing number of German politicians have suggested that conflict-related conditions should be attached to Nordstream 2, for example, after Russia detained Ukrainian boats in the Azov Sea in late 2018 (Chazan and Foy 2018).

While these efforts have patently failed to ensure harmony over Nordstream 2, a more robust consensus has taken root within the EU that a firmer set of responses is needed to deal with Russian geopolitical actions – and that the EU cannot afford to let divergence over energy policy cut across this higher-priority challenge. This geopolitical imperative has influenced the positions of market-oriented states like the Netherlands, Sweden and the UK, bringing them slightly more into line with the Commission and Eastern European states. Denmark is the one member state to have introduced a process that explicitly foregrounds the security implications of energy pipelines (cf. in this volume, Goldthau and Sitter 2020); in de facto terms other member states and the EU institutions increasingly follow something of the same logic. As of late 2019, an emerging debate is whether calls from President Macron for a cautious strategic rapprochement with Russia might now put the EU's fragile unity at risk (The Economist 2019).

A second example involves Iran. There has been contestation between member states over energy policy towards Iran for many years. Some member states – Italy being the prime example – have been keener than others to push ahead with new energy deals in Iran and with a more formalized EU-level energy partnership with Tehran. Increasingly, the EU has approached Iran as a priority geopolitical issue, seeking to get Iran to agree controls on its nuclear programme and also to influence Iran's actions in other parts of the Middle East. These overarching geopolitical concerns have taken precedence over other aspects of relations with Iran (European Commission and High Representative 2017; European Union 2015; Barnes-Dacey 2017; Koenig 2017; Smits et al. 2016; European Political Strategy Centre 2016). Convergence on the wider EU foreign-policy agenda towards Iran has helped soften previous differences between member states over the appropriateness of pursuing new energy deals in the country. Indeed, member states agreed on a Blocking Statute to protect European companies investing in Iran from US secondary sanctions as part of their attempt to save the Joint and Comprehensive Plan of Action after the Trump administration withdrew from the accord (Alcaro 2018).

These are only two examples of the geopolitical-energy spill-over – others could be given from across the Middle East and sub-Saharan Africa – but they suffice to demonstrate how high-priority strategic aims can encourage member states and EU institutions to keep contestation over energy policy within narrower limits than might otherwise be the case. From this flows a key reflection, relevant to the volume as a whole: debates over energy questions cannot be seen in isolation from the geopolitical challenges that increasingly crowd in on the union. Energy experts tend to broach foreign-policy issues in terms of the way that a priori energy goals spill-over to EU external relations. However, the inverse spill-over is also germane: that is, how the EU's highest priority and increasingly pressing geostrategic aims act as a limiting influence over energy-policy debates. This is not to suggest that these aims determine energy-policy decisions in any precise or primary sense, but rather to point out that they shape the context within which such decisions are made – and potentially in a way that leads to a bounded form of contestation. Somewhat contrary to received wisdom on this link, geopolitical constraints have on occasion acted at least partially and modestly to de-politicize EU energy policy.

VI. Conclusions

This article shows that a combination of internal and external factors is at play in the external elements of EU energy policy. These certainly drive sharp debates and differences between member states – this is the element that is best known and most extensively covered in the academic literature, including in this volume. Yet they also, at least to some extent, have limited contestation in recent years. This is where the chapter makes a distinctive contribution to the volume. In particular, it points out that factors exogenous to energy policy can be significant and how these can contain contestation as well as generate it.

The three policy strands covered in the article – security of supply, climate security and geopolitical factors – show a balance of contestation and convergence between different institutional actors. Member states and EU institutions still hold contrasting views on elements of external energy policy. However, these differences are overlain by areas of convergence, and where internal disagreements do exist they have in recent years often been kept within manageable proportions. The chapter supports the view that, while many actors have a role in EU energy policy and eschew any single vision or voice, they often pull in broadly similar directions (Richert 2017). It also resonates with suggestions that a 'lite' form of collective securitization has emerged where broad strategic concerns have pushed member states towards more coordination on external energy questions, even if this is not enough for national governments to support a truly united single energy policy (Hofmann and Staeger 2019).

Different logics contribute to this bounded contestation. The energy-policy divergence has been contained by several of the factors presented in the edition's overarching analytical framework: new EU mechanisms of horizontal coordination; strategies of balance and inter-issue compromise; a substantive convergence of the climate and security agendas; and the effect of exogenous crises in de-politicizing energy policy. Significantly, these logics chime with theoretical perspectives on EU foreign policy that detect an on-going shaping of common conceptions of security imperatives – showing that these academic frameworks have relevance for the more specific cross-over to energy policy.

In line with points made elsewhere in the volume (Goldthau and Sitter 2020), contestation in external energy policy is not primarily a search for formal control or a zero-sum battle over exactly where legal competences should lie. Rather, the conferral of authority is reshaped when institutions gain de facto policy legitimacy and help infuse some degree of EU strategic commonality. Internal EU debate is framed mainly in terms of the precise balance to be struck between the triangle of internal market logics, external security of supply and support for energy transition around the world – with exogenous security crises adding another ingredient. The EU foreign-policy community has gained a stronger toehold in energy policy, but has broadly accepted the primary importance of climate imperatives. This chapter supports others' findings that the substantive focus on the security of supply has become more notable relative to the reliance on purely market dynamics (Goldthau and Sitter 2020). It also shows, however, there are climate-policy developments and wider geopolitical factors that have become additionally influential in tandem with this shift.

Linking these diverse trends together, it might be said that the overarching logic is one of the EU intricately balancing security of supply, climate policies and other, non-energy security objectives. This accords with what some have stressed is an increasingly determinant triangular nexus of policy concerns involving energy markets, geopolitics and climate sustainability (Herranz-Surrallés 2018). The article suggests that – at least to a modest extent – this emerging triangle is helping to soften some of the roughest edges of internal EU contestation.

Disclosure statement

No potential conflict of interest was reported by the author.

References

Aggestam, L., and F. Bicchi. 2019. "New Directions in EU Foreign Policy Governance: Cross-Loading, Leadership and Informal Groupings." *Journal of Common Market Studies* 57/3: 515–532. doi:10.1111/jcms.12846.

Alcaro, R. 2018. *Europe and Iran's Nuclear Crisis: Lead Groups and EU Foreign Policy-making*. London: Palgrave Macmillan.

Amadio Viceré, M. G. 2016. "The Roles of the President of the European Council and the High Representative in Leading EU Foreign Policy on Kosovo." *Journal of European Integration* 38 (5): 557–570. doi:10.1080/07036337.2016.1178255.

Andoura, S., and J.-A. Vinois. 2015. *From the European Energy Community to the Energy Union: A Policy Proposal*. Paris: Notre Europe Jacques Delors Institute.

Arias Cañete, M. 2018 April 12. *Speech at the 4th EU Energy Summit: International Geopolitical Uncertainties: Brakes or Accelerators for the EU Energy Transition?* Brussels: European Commission.

Barnes-Dacey, J. 2017. *To End a War: Europe's Role in Bringing Peace in Syria*. London: European Council on Foreign Relations.

Bergamaschi, L., and N. Sartori. 2018. *The Geopolitics of Climate: Transatlantic Dialogue*. Rome: Istituto Affari Internazionali.

Bickerton, C., D. Hodson, and U. Puetter. 2015. "The New Intergovernmentalism: European Integration in the post-Maastricht Era." *Journal of Common Market Studies* 53: 4. doi:10.1111/jcms.12212.

Birchfield, V. 2013. "A Normative Power Europe Framework of Transnational Policy Formation." *Journal of European Public Policy* 20 (6): 907–922. doi:10.1080/13501763.2013.781829.

Birchfield, V., and J. Duffield, eds. 2011. *Toward a Common European Union Energy Policy*. London: Palgrave Macmillan.

Biscop, S. 2015. *Global and Operational: A New Strategy for EU Foreign and Security Policy*. IAI Working Papers 15/27. Rome: Istituto Affari Internazionali.

Borrell, J. 2019. "Confirmation Hearing." *European Parliament* 7 (October): 2019.

Börzel, T., and V. van Hüllen. 2015. "One Voice, One Message, but Conflicting Goals: Cohesiveness and Consistency in the European Neighbourhood Policy." *Journal of European Public Policy* 21 (7): 1033–1049. doi:10.1080/13501763.2014.912147.

BP. 2018. "BP Energy Outlook 2018." Accessed October 2018. https://www.bp.com/content/dam/bp/en/corporate/pdf/energy-economics/energy-outlook/bp-energy-outlook-2018.pdf

Buchan, D., and M. Keay. 2015. *Europe's Energy Union Plan: A Reasonable Start to a Long Journey*. Oxford Energy Comment, Oxford: The Oxford Institute for Energy Studies.

Chazan, G., and H. Foy 2018. "Frontrunners to Succeed Merkel Raise Questions over Russian Pipeline." *Financial Times*, December 4.

Climate Action. 2018. "No Paris Agreement, No EU Trade Deal' Says France to US." February 5. Accessed February 2019. http://www.climateaction.org/news/no-paris-agreement-no-eu-trade-deal-says-france-to-us

Costa, O. 2018. "The Politicization of EU External Relations." *Journal of European Public Policy* 26 (5): 790–802.

Council of the EU. 2015. *Council Conclusions on EU Energy Diplomacy*. Brussels: Council of the EU.

Council of the EU. 2018. "Foreign Affairs Council Conclusions on Climate Diplomacy." 6125/18, February 26.

Council of the EU. 2019. *Council Conclusions on Climate Diplomacy*. 6153/19.

De Jong, D. 2016. "Why Europe Should Fight Nord Stream II." *EU Observer*, February 23.

Delcour, L., and H. Kostanyan. 2014. *Towards a Fragmented Neighbourhood: Policies of the EU and Russia and Their Consequences for the Area that Lies in Between*. Brussels: Centre for European Policy Studies.

Escribano, G. 2019. "The Geopolitics of Renewable and Electricity Cooperation between Morocco and Spain." *Mediterranean Politics* 24 (5): 674–681. doi:10.1080/13629395.2018.1443772.

European Commission. 2015. *A Framework Strategy for A Resilient Energy Union with A Forward-Looking Climate Change Policy*. COM(2015) 80 final, 25 February.

European Commission. 2017. "Third Report on the State of the Energy Union." COM(2017) 688 final, November 23.

European Commission. 2018. "A Clean Planet for All: A European Strategic Long-term Vision for A Prosperous, Modern, Competitive and Climate Neutral Economy." COM(2018) 773, November 28.

European Commission. 2019. "International Climate Finance." Accessed March 2019. https://ec. europa.eu/clima/policies/international/finance_en

European Commission. 2019b. "The European Green Deal." COM(2019)640.

European Commission and High Representative for Foreign Affairs and Security Policy. 2017. "*Elements for an EU Strategy for Syria.*" JOIN(2017)11, March 14.

European External Action Service. 2019. *The European Union's Global Strategy: Three Years On, Looking Forward*. Brussels: European External Action Service.

European Parliament. 2017. *Intergovernmental Agreements in the Field of Energy*. EPRS Briefing.

European Political Strategy Centre. 2016. *Towards a 'security Union': Bolstering the EU's Counter-Terrorism Response*. Brussels: European Political Strategy Centre.

European Union. 2015. "Strategic Orientation Document for the EU Regional Trust Fund in Response to the Syrian Crisis." Accessed March 2019. https://ec.europa.eu/trustfund-syria-region/sites/tfsr/files/eu_regional_tf_madad_syrian_crisis_strategic_orientation_paper1.pdf

European Union. 2016. "Shared Vision, Common Action: A Stronger Europe, A Global Strategy for the European Union's Foreign and Security Policy." (Accessed March 2019. http://eeas.europa.eu/archives/docs/top_stories/pdf/eugs_review_web.pdf

Far, S., and R. Youngs. 2015. *Energy Union and EU Global Strategy*. Stockholm: Swedish Institute for European Policy Studies.

Fetzek, S., and L. van Schaik. 2018. *Europe's Responsibility to Prepare: Managing Climate Security Risks in a Changing World*. Washington DC: Center on Climate and Security.

Fischer, S., and O. Geden. 2015. "The Limits of "Energy Union"." SWP Comments.

Goldthau, A., and N. Sitter. 2015a. *A Liberal Actor in A Realist World. The European Union Regulatory State and the Global Political Economy of Energy*. Oxford: Oxford University Press.

Goldthau, A., and N. Sitter. 2015b. "Soft Power with a Hard Edge: EU Policy Tools and Energy Security." *Review of International Political Economy* 22: 941–965. doi:10.1080/09692290.2015.1008547.

Goldthau, A., and N. Sitter. 2020. "Power, Authority and Security: The Eu's Russian Gas Dilemma." *Journal of European Integration* 42 (1).

Gstöhl, S., and S. Schunz, eds. 2016. *Theorizing the European Neighbourhood Policy*. London: Routledge.

Haukkala, H. 2015. "From Cooperative to Contested Europe? The Conflict in Ukraine as a Culmination of a Long-Term Crisis in EU–Russia Relations." *Journal of Contemporary European Studies* 23 (1): 25–40. doi:10.1080/14782804.2014.1001822.

Helm, D. 2015. *The Energy Union: More than the Sum of Its Parts?* London: Centre for European Reform.

Herranz-Surrallés, A. 2016. "An Emerging EU Energy Diplomacy? Discursive Shifts, Enduring Practices." *Journal of European Public Policy* 23 (9): 1386–1405. doi:10.1080/13501763.2015.1083044.

Herranz-Surrallés, A. 2018. "Thinking Energy outside the Frame? Reframing and Misframing in Euro-Mediterranean Energy Relations." *Mediterranean Politics* 23 (1): 122–141. doi:10.1080/13629395.2017.1358903.

Herranz-Surrallés, A., I. Solorio, and J. Fairbrass. 2020. "Renegotiation Authority in The Energy Union: A Framework for Analysis." *Journal of European Integration* 42 (1).

Hill, C., M. Smith, and S. Vanhoonacker. 2017. *International Relations and the European Union*. Third ed. Oxford: Oxford University Press.

Hofmann, S., and U. Staeger. 2019. "Frame Contestation and Collective Securitisation: The Case of EU Energy Policy." *West European Politics* 42 (2): 323–345. doi:10.1080/01402382.2018.1510197.

Hyde-Price, A. 2007. *European Security in the 21st Century: The Challenge of Multipolarity*. London: Routledge.

Juncos, A., and K. Pomorska. 2015. "Attitudes, Identities and an Emergence of an Esprit De Corps in the EEAS." In *The European External Action Service. European Diplomacy Post-Westphalia*, edited by J. Batora and D. Spence, 373–391. London: Palgrave Macmillan.

Keukelaire, S., and T. Delreux. 2014. *The Foreign Policy of the European Union*. Second ed. Basingstoke: Palgrave Macmillan.

Koenig, N. 2017. "Libya and Syria: Inserting the European Neighbourhood Policy in the European Union's Crisis Response Cycle." *European Foreign Affairs Review* 22 (1): 19–38.

Krotz, U., and R. Maher. 2012. "Debating the Sources and Prospects of European Integration." *International Security* 37 (1): 178–199. doi:10.1162/ISEC_c_00092.

Lavenex, S. 2014. "The Power of Functionalist Extension: How EU Rules Travel." *Journal of European Public Policy* 21 (6): 885–903. doi:10.1080/13501763.2014.910818.

Lucarelli, S. 2019. "The EU as a Securitising Agent? Testing the Model, Advancing the Literature." *West European Politics* 42 (2): 413–436. doi:10.1080/01402382.2018.1510201.

Manners, I. 2008. "The Normative Ethics of the European Union." *International Affairs* 84 (1): 45–60. doi:10.1111/inta.2008.84.issue-1.

Morillas, P. 2019. "Autonomy in Intergovernmentalism: The Role of De Novo Bodies in External Action during the Making of the EU Global Strategy." *Journal of European Integration* 1–16. doi:10.1080/07036337.2019.1666116.

New Europe. 2019. "MEPs Propose that 45% of NDICI Funds Should Support Climate and Environmental Objectives." March 7. Accessed March 2019. https://www.neweurope.eu/article/eu-funding-to-support-climate-and-environmental-goals/

New Europe. 2019b. "EIB to Support Climate Action Projects in Brazil." 27 October 2019.

New Europe. 2019c. "EU Says Russia Gas Transit Talks through Ukraine Hit Skids." 2 November 2019.

Nitoiu, C., ed. 2016. *Avoiding a New 'cold War': The Future of EU-Russia Relations in the Context of the Ukraine Crisis*. London: London School of Economics.

Nixey, J. 2014. *Russia and the EU are Signing Their Divorce Papers*. London: Chatham House.

Raik, K. 2016. "Liberalism and Geopolitics in EU–Russia Relations: Rereading the Baltic Factor." *European Security* 25 (2): 237–255. doi:10.1080/09662839.2016.1179628.

Richert, J. 2017. "From Single Voice to Coordinated Polyphony: EU Energy Policy and the External Dimension." *European Foreign Affairs Review* 22 (2): 213–232.

Rosato, S. 2011. "Europe's Troubles: Power Politics and the State of the European Project." *International Security* 35 (4): 45–86. doi:10.1162/ISEC_a_00035.

Rutte, M. 2019. "Churchill Lecture by Prime Minister Mark Rutte, Europa Institut at the University of Zurich." Accessed March 2019. https://www.government.nl/documents/speeches/2019/02/13/churchill-lecture-by-prime-minister-mark-rutte-europa-institut-at-the-university-of-zurich

Sakwa, R. 2015. *Frontline Ukraine: Crisis in the Borderlands*. London: I. B. Tauris.

Schimmelfennig, F. 2015. "What's the News in "New Intergovernmentalism"? A Critique of Bickerton, Hodson and Puetter." *Journal of Common Market Studies* 53(4): 723–730.

Schmidt, V. A. 2010. "Taking Ideas and Discourse Seriously: Explaining Change through Discursive Institutionalism as the Fourth New Institutionalism." *European Political Science Review* 2 (1): 1–25. doi:10.1017/S17557739099901X.

Schmidt-Felzmann, A. 2020. "Nord Stream 2 and Diffuse Authority in The Eu: Managing Authority Challenges regarding Russian Pipelines in the Baltic Sea Area." *Journal of European Integration* 42 (1).

Scholten, D., I. Ydersbond, T. Sattich, and T. H. Inderberg. 2015. *Consensus, Contradiction, and Conciliation of Interests: The Geo-economics of the Energy Union*. Brussels: European Policy Centre.

Schunz, S. 2012. "Explaining the Evolution of European Union Foreign Climate Policy: A Case of Bounded Adaptiveness." *European Integration Online Papers* 16 (article): 6.

Siddi, M. 2019. "The EU's Botched Geopolitical Approach to External Energy Policy: The Case of the Southern Gas Corridor." *Geopolitics* 24 (1): 124–144. doi:10.1080/14650045.2017.1416606.

Smits, R., F. Molenaar, F. El-Kamouni-Janssen, and N. Grinstead. 2016. *Cultivating Conflict and Violence? A Conflict Perspective on the EU Approach to the Syrian Refugee Crisis*. The Hague: Clingendael.

Sperling, J., and M. Webber. 2019. "The European Union: Security Governance and Collective Securitisation." *West European Politics* 42 (2): 228–260. doi:10.1080/01402382.2018.1510193.

Szulecki, K., and K. Westphal. 2014. "The Cardinal Sins of European Energy Policy: Non-Governance in an Uncertain Global Landscape." *Global Policy* 5 (1): 38–51. doi:10.1111/gpol.2014.5.issue-s1.

Tèlo, M. 2007. *Europe: A Civilian Power? European Union, Global Governance, World Order*. Basingstoke: Palgrave, Macmillan.

The Economist. 2019. "Transcript of Interview with President Macron", November 7.

Thomas, D., ed. 2011. *Making EU Foreign Policy*. Basingstoke: Macmillan.

Toplensky, R. 2019. "EU Finalises Tougher Rules on $9.5bn Russia-Germany Gas Pipeline." *Financial Times*, February 13. Accessed March 2019. https://www.neweurope.eu/article/eu-funding-to-support-climate-and-environmental-goals/

Van der Linde, C. 2007. "External Energy Policy: Old Fears and New Dilemmas in a Larger Union." In *Fragmented Power: Europe and the World Economy*, edited by A. Sapir, 266–307. Brussels: Bruegel.

Wivel, A., and O. Wæver. 2018. "The Power of Peaceful Change: The Crisis of the European Union and the Rebalancing of Europe's Regional Order." *International Studies Review* 20: 317–325. doi:10.1093/isr/viy027.

Wright, T. 2012. "What If Europe Fails?" *Washington Quarterly* 35 (3): 23–41. doi:10.1080/0163660X.2012.703584.

Youngs, R. 2014. *European Security and Climate Change*. London: Routledge.

Youngs, R. 2017. *Europe's Eastern Crisis*. Cambridge: Cambridge University Press. doi:10.1017/9781316344033.

Index